Foreign Dollar Balances and the International Role of the Dollar

NATIONAL BUREAU OF ECONOMIC RESEARCH

Studies in International Economic Relations

1. Problems of the United States as World Trader and Banker *Hal B. Lary*

2. Price and Quantity Trends in the Foreign Trade of the United States *Robert E. Lipsey*

3. Measuring Transactions Between World Areas *Herbert B. Woolley*

4. Imports of Manufactures from Less Developed Countries *Hal B. Lary*

5. The Responsiveness of Demand Policies to Balance of Payments: Postwar Patterns *Michael Michaely*

6. Price Competitiveness in World Trade *Irving B. Kravis and Robert E. Lipsey*

7. Foreign Dollar Balances and the International Role of the Dollar *Raymond F. Mikesell and J. Herbert Furth*

8. Foreign Trade Regimes and Economic Development, Vols. 1–11 *Jagdish N. Bhagwati and Anne O. Krueger,* editors

Foreign Dollar Balances and the International Role of the Dollar

RAYMOND F. MIKESELL

UNIVERSITY OF OREGON

and

J. HERBERT FURTH

FACULTY ASSOCIATE EMERITUS,

FOREIGN SERVICE INSTITUTE

NATIONAL BUREAU OF ECONOMIC RESEARCH

NEW YORK

1974

Distributed by COLUMBIA UNIVERSITY PRESS

NEW YORK AND LONDON

HG
3881
.M514

Printed in the United States of America

V

John F. Kain	Robert E. Lipsey	Ilse Mintz	Robert P. Shay
John W. Kendrick	Sherman J. Maisel	Geoffrey H. Moore	Edward K. Smith
Irving B. Kravis	Benoit B. Mandelbrot	M. Ishaq Nadiri	George J. Stigler
Edwin Kuh	John R. Meyer	Nancy Ruggles	Victor Zarnowitz
William M. Landes	Robert T. Michael	Richard Ruggles	
Hal B. Lary	Jacob Mincer	Anna J. Schwartz	

Relation of the Directors to the Work and Publications
of the National Bureau of Economic Research

1. The object of the National Bureau of Economic Research is to ascertain and to present to the public important economic facts and their interpretation in a scientific and impartial manner. The Board of Directors is charged with the responsibility of ensuring that the work of the National Bureau is carried on in strict conformity with this object.

2. The President of the National Bureau shall submit to the Board of Directors, or to its Executive Committee, for their formal adoption all specific proposals for research to be instituted.

3. No research report shall be published until the President shall have submitted to each member of the Board the manuscript proposed for publication, and such information as will, in his opinion and in the opinion of the author, serve to determine the suitability of the report for publication in accordance with the principles of the National Bureau. Each manuscript shall contain a summary drawing attention to the nature and treatment of the problem studied, the character of the data and their utilization in the report, and the main conclusions reached.

4. For each manuscript so submitted, a special committee of the Board shall be appointed by majority agreement of the President and Vice-Presidents (or by the Executive Committee in case of inability to decide on the part of the President and Vice-Presidents), consisting of three directors selected as nearly as may be one from each general division of the Board. The names of the special manuscript committee shall be stated to each Director when the manuscript is submitted to him. It shall be the duty of each member of the special manuscript committee to read the manuscript. If each member of the manuscript committee signifies his approval within thirty days of the transmittal of the manuscript, the report may be published. If at the end of that period any member of the manuscript committee withholds his approval, the President shall then notify each member of the Board, requesting approval or disapproval of publication, and thirty days additional shall be granted for this purpose. The manuscript shall then not be published unless at least a majority of the entire Board who shall have voted on the proposal within the time fixed for the receipt of votes shall have approved.

5. No manuscript may be published, though approved by each member of the special manuscript committee, until forty-five days have elapsed from the transmittal of the report in manuscript form. The interval is allowed for the receipt of any memorandum of dissent or reservation, together with a brief statement of his reasons, that any member may wish to express; and such memorandum of dissent or reservation shall be published with the manuscript if he so desires. Publication does not, however, imply that each member of the Board has read the manuscript, or that either members of the Board in general or the special committee have passed on its validity in every detail.

6. Publications of the National Bureau issued for informational purposes concerning the work of the Bureau and its staff, or issued to inform the public of activities of Bureau staff, and volumes issued as a result of various conferences involving the National Bureau shall contain a special disclaimer noting that such publication has not passed through the normal review procedures required in this resolution. The Executive Committee of the Board is charged with review of all such publications from time to time to ensure that they do not take on the char-

vii

acter of formal research reports of the National Bureau, requiring formal Board approval.

7. Unless otherwise determined by the Board or exempted by the terms of paragraph 6, a copy of this resolution shall be printed in each National Bureau publication.

*(Resolution adopted October 25, 1926, and revised February 8, 1933,
February 24, 1941, and April 20, 1968)*

Preface

This study was initiated by J. Herbert Furth several years before the international financial crises which led to the termination of the gold convertibility of the dollar and to the current effort of the members of the International Monetary Fund to reform the international monetary system. It has been difficult both to delimit and terminate this study (which has gone through many drafts) because the rapid developments in international finance have continually brought new problems to the foreground while reducing the significance of others. The study is concerned with many of the same problems dealt with by Hal B. Lary in his book *Problems of the United States as World Trader and Banker* (NBER, 1963), but our empirical analysis has focused on the origin, composition, and behavior of foreign holdings of liquid dollar assets, including both assets held in the United States and Eurodollars. In addition to the statistical analysis of the behavior of foreign dollar holdings, the present study sets forth an analytical framework for studying the interrelations among the several types of liquid dollar assets as determined by the behavior of the major categories of foreign portfolio holders. In our concluding chapter we have sought to relate our findings to the problem of creating a more satisfactory system of international monetary reserves and balance of payments adjustment.

This study forms part of the National Bureau's continuing program of research on international trade and payments problems. Previous studies in this series have included Hal B. Lary's books on *Problems of the United States as World Trader and Banker* (NBER, 1963) and on *Imports of Manufactures from Less Devel-*

oped Countries (CUP, 1968); *Price and Quantity Trends in the Foreign Trade of the United States,* Robert E. Lipsey (PUP, 1963); *Measuring Transactions Between World Areas,* Herbert B. Woolley (CUP, 1966); *The Responsiveness of Demand Policies to Balance of Payments: Postwar Patterns,* Michael Michaely (CUP, 1971); *Price Competitiveness in World Trade,* Irving B. Kravis and Robert E. Lipsey (CUP, 1971); and various volumes on the subject: *Foreign Trade Regimes and Economic Development,* Jagdish N. Bhagwati and Anne O. Krueger, editors.

The study has been financed in part by a grant from the Ford Foundation and in part by the National Bureau. Financing for secretarial and statistical services, computer time, and office materials as well as for a substantial portion of the time of one of the authors has been provided by the University of Oregon.

A number of individuals have given generously of their time in reading the manuscript at various stages and in providing information.

The following economists have made valuable suggestions for the improvement of the manuscript without necessarily agreeing with all of the analysis: Philip Cagan, Peter B. Kenen, Fred H. Klopstock, Miroslav Kriz, Walter Lederer, Robert L. Sammons, and Carl H. Stem. We are grateful for helpful comments and criticism received from the Board of Directors reading committee, consisting of Eugene A. Birnbaum, Robert V. Roosa, and Willard L. Thorp.

The authors also want to acknowledge the assistance of the following individuals in providing unpublished data and other information necessary for this study: C. L. Callander, Robert F. Gemmill, Helmut Mayer, Phillip E. Schaffner and Wilson E. Schmidt. Finally, the authors owe a special debt of gratitude to Hal B. Lary who provided constant guidance and made numerous substantive and editorial comments and suggestions during the preparation of this manuscript.

Contents

Tables

Figures

Foreign Dollar Balances and the International Role of the Dollar

1

Background and Objectives

Origins of the Study

WORK on this study began several years ago in response to two somewhat conflicting trends in thinking about international monetary and payments problems. There was, on the one hand, a widely held view that the growth of international liquidity—freely convertible and universally acceptable means of payment for transacting international business and settling balances between countries —was not assured in sufficient size and regularity to serve the needs of international trade and the orderly expansion of the world economy. On the other hand, there was a growing apprehension that the additional liquidity being generated consisted too largely of dollars and was too dependent on continuing deficits in the U.S. balance of payments as a means of supplying additional dollar balances to foreigners.

There was perhaps a certain contradiction in these two views. If the addition to foreign holdings of dollars was not excessive in relation to the presumed secular rise in international liquidity requirements, there need not have been serious grounds for worry about the chronic U.S. deficit. In the extreme view,[1] however, there was no compromise: to meet liquidity needs solely or mainly by increasing foreign dollar balances presupposed a chronic payments deficit by the United States such as to weaken confidence in the dollar and undermine the system. Coupled with this view was the belief that the system as it operated was easily subject to abuse, that it did not impose discipline on the United States comparable to

1. As expressed notably by Robert Triffin in *Gold and the Dollar Crisis,* New Haven: Yale University Press, 1960.

1

that faced by other countries in deficit, and that the amount of dollars which foreign central banks found themselves compelled to accept could, and probably would, exceed their needs and wishes.

Under these conditions, there was a need to know more about the composition, motivation, and growth prospects of foreign holdings of liquid dollar balances. The question refers particularly, of course, to foreign private holdings, since central banks were obliged under the prevailing system to absorb any dollars not otherwise taken up on the foreign exchange market.

What, then, were the reasons foreign commercial banks, business enterprises, and individuals chose to hold dollars? What purposes did they serve? How much growth might reasonably be expected? How firmly were the dollars held or, conversely, how much of a threat did the mounting volume of foreign liquid claims pose to the ability of the United States to defend the exchange value of the dollar?

In brief, questions of this sort presupposed an analysis in terms of what has come to be called a portfolio adjustment model. What were the preference functions of foreign holders of dollars? What were the key variables? How did the variables move?

Along with the buildup of foreign liquid claims on the United States came the curious phenomenon of the rapid growth of the Eurodollar market—a market for borrowing and lending American dollars operated outside the United States by foreign banks, including foreign branches of U.S. banks. What were the reasons for this development? How did the Eurodollar relate to the American dollar in terms of the functions served by both in international trade and finance? How did the Eurodollar market relate to flows of liquid funds to and from the United States? Was the Eurodollar market in some sense a part of the U.S. payments problem? Did it mitigate or accentuate the problem?

The original aims of this study, initiated by J. Herbert Furth in 1967, have been overtaken by events, culminating in the confidence crisis in the spring of 1971, the formal suspension of convertibility of the dollar in August 1971, and the devaluation of the dollar and revaluation of other leading currencies in December 1971, and the further depreciation of the dollar in February 1973. It is worth noting that, when the crunch came, American-owned dollars proved just as flight-prone as foreign-owned dollars, to judge by the size of unrecorded capital outflows as mirrored in the errors and omissions item of the U.S. balance of payments. As might have been known from confidence crises that have afflicted other currencies in earlier times, this recent experience suggests that countries in basic payments disequilibrium have to be mindful not only of their liabilities to, and claims on, other countries but of the potential mobility of domestic funds as well.[2]

Though affected by these recent developments, the original aims of the study nevertheless remain relevant to the evolving world monetary and payments system. They are pertinent to an understanding of the past and to an evaluation of future problems and prospects. Unfortunately, we are still far from being able to provide the answers, or to command the data, required of a properly developed set of portfolio adjustment models. We can go only part way, throw some light on key issues, and help to frame the questions on which further research is needed.

Questions to Be Considered

The principal subjects dealt with in this report fall into three categories, explored in the following three chapters respectively.

First, in Chapter 2, we have found it necessary to review and to some extent reconstruct the statistical record with respect to foreign holdings of dollars. As far as foreign balances in the United States —i.e., American dollars—are concerned, we can be relatively brief. This is particularly true inasmuch as the U.S. Department of Commerce has recently modified the composition and arrangement of its figures along lines that seemed essential to us when this study was

---

2. This was, in fact, a major weakness in the "liquidity balance" used by the Department of Commerce for many years as the principle measure of balance in the U.S. international accounts. It regarded increases in both private and official liquid dollar assets by foreigners as a potential threat to the exchange value of the dollar. Consequently according to the "liquidity balance," all increases in foreign holdings of liquid dollar assets tended to increase the U.S. deficit and vice versa. Logically, however, both U.S. resident and nonresident dollar holdings can be exchanged for foreign currencies in the exchange market. Hence, whether an international transaction increases or decreases the vulnerability of the dollar to depreciation in the exchange market is not a proper criterion for the selection of an accounting balance for measuring equilibrium.

undertaken. Now the criteria and presentation seem to be sufficiently close to our own views so that only minor modifications in the Department of Commerce format are made here. Even so, we have found it essential to bring into consideration other highly relevant data, hitherto little used, on the role of multinational banks and the important part they play in this country's dollar liabilities. We have in mind both the internal obligations of U.S. parent banks to their foreign branches and those of the U.S. branches and agencies of foreign banks to their home offices abroad.

Also in Chapter 2 we explore the statistical intricacies of the Eurodollar market. Since deposits in that market are not claims on U.S. residents, one may ask if they really need to be taken into account. We argue that they should be, given the close similarity in some of the important functions served by the Eurodollar and the American dollar. We then trace the development of Eurodollar deposits, by main categories of holders, on the basis of statistics rather different from those made familiar in the annual reports of the Bank for International Settlements.

In Chapter 3 we analyze the behavior of foreign dollar holdings by the three major categories of foreign holders, namely, private nonbanks, commercial banks and official institutions. We begin with a brief discussion of the origins of foreign dollar balances, which includes our own contribution to the controversial literature on the generation of Eurodollar deposits. In line with the portfolio-adjustment approach, we have sought to determine the factors that govern the preference functions of foreign private liquid asset holders, which in turn determine the demand for American dollars, Eurodollars, and other currencies. Within the limits of our data we examine the sensitivity of foreign nonbank holdings of U.S. dollars, Eurodollars, and foreign currencies to changes in interest rate differentials corresponding to the three categories of liquid assets. Our modest results suggest that further analysis may prove rewarding, especially if certain data limitations can be overcome. In this chapter we also deal with the interrelationships between the operations of the Eurodollar market and the U.S. balance of payments. The interrelations between the Eurodollar market and U.S. short-term capital movements are strong and subject to quantitative measurement. But the effects of the market on the U.S. basic balance can only be conjectured.

In our concluding chapter we summarize our empirical findings and apply them to certain basic issues relating to the past and possible future international role of the dollar. Since a substantial portion of the international financial intermediation functions of U.S. financial markets has been transferred to the Eurocurrency and Eurobond markets, we explore the implications of the development of these Euromarkets for the international role of the dollar and for the U.S. balance of payments. Finally, we apply our empirical and analytical findings to the problems of international monetary reform currently under negotiation by the Committee of Twenty. In spite of the recent dollar crises, we find that the current position and functions of the dollar may prove difficult, if not impossible, to replace, and may impose an insurmountable barrier to monetary authorities seeking to make the dollar symmetrical with other major currencies in the world payments system.

2

Foreign Dollar Balances: A Review and Extension of the Statistical Record

A Functional Approach

THE principal aim of this chapter is to provide a summary view of the growth of foreign holdings of dollar balances in recent years. We shall look first at foreign liquid claims on the United States on both a gross and net basis; then at liquid dollar claims by foreigners on each other in the Eurodollar market, again on a gross and net basis; and finally at the picture which emerges when foreign holdings of American dollars and Eurodollars are combined.

The selection, classification, and combination of items entering into the statistics are necessarily related to the functions which the dollar performs in the international economy. It is suggested that the principal functions are as follows:

1. To provide foreign central banks and other official institutions with a medium for holding reserves.

2. To provide an international standard of value and an intervention currency for maintaining the desired pattern of exchange rates.

3. To provide foreign commercial banks and businesses with transactions balances and precautionary balances for use in effecting payments in international trade.

4. To provide a medium for financing capital transactions.

5. To provide foreign commercial banks, business firms, and individuals with an attractive medium for investing in interest-earning assets of relatively high liquidity.

6

Consideration of these functions is helpful in deciding what to include in compiling statistics on liquid dollar claims whether these claims are those of foreigners on U.S. residents, or those of U.S. residents on foreigners, or Eurodollar claims involving claims of foreigners on other foreigners. For some purposes it would be desirable to include in liquid dollar claims all short-term claims calling for payment of dollars.[1] However, data are lacking for measuring the total volume of short-term dollar claims, and, in addition, definitions of liquid dollar liabilities and claims which depart substantially from those employed in the compilation of U.S. official statistics would be confusing. Hence, with minor exceptions, we shall include in U.S. liquid liabilities to foreigners those items regarded as liquid in Department of Commerce balance-of-payments statistics. These items with minor changes are detailed in an appendix to this chapter.

We are also concerned with liquid claims of U.S. residents on foreigners. These data enable us to analyze the net international liquidity position of the United States. For some purposes it is also instructive to examine the net liquid dollar position of major categories of foreigners vis-à-vis U.S. residents.[2] Our data include those items listed as U.S. liquid claims on foreigners in U.S. official statistics plus certain additional items. (See Appendix A to this chapter.)

Our functional approach has also led us to consider the role of Eurodollars and to compile Eurodollar statistics from the various sets of partial data available. From the point of view of analyzing the U.S. balance-of-payments position, Eurodollars are, of course, not claims on the United States. Nevertheless we must recognize

1. Liquid claims as defined by the Department of Commerce consist mainly of short-term claims on banks, U.S. government securities (both short-term and long-term) and marketable short-term claims on nonbanks. Short-term claims on foreign commercial firms arising out of international trade transactions are not regarded as liquid. (For a list of liquid claims according to the Department of Commerce definition, see Appendix A to this chapter.)

2. It would also be useful to examine the net liquid dollar position of major categories of foreigners vis-à-vis U.S. residents by major country or region since, for example, nonbank net creditors may be located in one region and nonbank net debtors in another. Unfortunately, however, data are not available to the authors for this type of disaggregation.

that Eurodollars serve some of the same functions as American dollars—particularly functions (1) and (5) above—and are directly redeemable in American dollars. This subject will be taken up in a separate section of this chapter.

In our statistical analysis of liquid dollar claims we have employed three categories of foreign dollar holders, namely, official institutions, commercial banks, and private nonbanks. Although each of these categories of foreign holders employs dollar balances for more than one purpose, each category of transactor can be assumed to have a unique demand function for dollar balances. Thus the demand for liquid dollar assets by *foreign official institutions* is in large measure a demand for official reserves held in the form of dollars. This demand function is a complex one in the case of the large developed countries since, in recent years, the volume of their official dollar holdings has reflected in considerable measure their exchange rate policies, i.e., the acquisition of dollars in order to avoid an appreciation of their currencies in relation to the dollar. *Foreign commercial bank* demand for American dollars and Eurodollars largely reflects their financial operations as lenders and borrowers (i.e., recipients of deposits) in a variety of currencies. The rapid changes in liquid dollar assets and liabilities of foreign commercial banks in recent years can only be understood through an analysis of the operations of the Eurodollar market. Since a large portion of the recorded liquid dollar claims constitutes internal accounting entries of multinational banks, special consideration will be given to this type of dollar claim. Finally, *foreign nonbanks* demand dollar balances to finance international transactions and to provide them with a highly liquid investment media. While the first function must be served by American dollars, in recent years the second function has been served more by Eurodollar deposits.

Foreign Holdings of American Dollars

Table 2.1 summarizes the changes in foreign liquid claims on U.S. residents between 1957 and 1971[3] by category of holder, and in

3. The statistics employed in this chapter do not extend beyond the end of 1971, but preliminary figures for the end of 1972 are given in the Appendix Tables.

C. Other foreigners										
1. Demand deposits in U.S. banks	1.8	1.5	1.5	1.6	1.5	1.7	1.8	1.7	1.7	1.7
2. Short-term time deposits in U.S. banks		1.0	1.3	1.6	1.8	2.1	2.2	1.9	1.9	1.7
3. Other liquid claims	0.9	0.9	1.0	1.0	1.0	1.0	1.0	1.0	1.1	0.8
Subtotal	2.7	3.4	3.8	4.1	4.3	4.7	5.0	4.6	4.7	4.2
Total	14.9	23.6	26.9	27.3	29.2	34.0	36.9	44.4	46.0	66.2
II. U.S. liquid claims on foreigners										
A. Central banks and other official agencies	0.2	0.4	0.7	1.1	1.6	2.7	3.8	3.0	0.7	0.5
B. Commercial banks, including U.S. branches	1.2ᵇ	4.6	6.2	5.6	5.9	6.4	6.9	7.4	7.7	9.9
C. Other foreigners	0.0ᵇ	0.2ᵇ	0.6	0.2	0.2	0.3	0.2	0.3	0.4	0.4
Total	1.4ᵇ	5.1ᵇ	7.5	6.9	7.7	9.4	10.9	10.7	8.8	10.8
III. Internt'l. liquidity position of the U.S.										
A. Central banks and other official agencies	−8.5	−14.0	−15.1	−14.7	−13.3	−15.5	−13.5	−13.0	−23.1	−50.2
B. Commercial banks, incl. U.S. branches	−2.3ᵇ	−1.2	−1.1	−1.8	−4.1	−4.7	−7.7	−16.4	−9.8	−1.4
C. Other foreigners	−2.7ᵇ	−3.2ᵇ	−3.2	−3.9	−4.1	−4.4	−4.8	−4.3	−4.3	−3.8
Net position with foreigners (A, B and C)	−13.5	−18.4	−19.4	−20.4	−21.5	−24.6	−26.0	−33.7	−37.2	−55.4
U.S. reserves (gold, SDRs, and IMF gold tranche)	24.6	15.8	15.4	13.8	12.6	11.5	11.2	13.2	13.3	11.3
Net position on all accounts combined	11.1	−2.6	−4.0	−6.6	−8.9	−13.1	−14.8	−20.5	−23.9	−44.1

SOURCES: Appendix tables.

NOTE: Columns may not add due to rounding.

a. Estimates for end of 1968, 1969, 1970, and 1971 from *Survey of Current Business*; estimates for earlier years from *Federal Reserve Bulletin*.

b. Excludes liquid claims reported by nonbanking concerns.

c. Change in reporting coverage. Figures in parentheses are comparable in coverage with those on preceding date.

at the end of 1957 to a high of $5.0 billion at the end of 1968 and then declined to $4.2 billion by the end of 1971. The behavior of these claims may well have been influenced by the growth of the Eurodollar market and the more attractive yields on Eurodollar deposits as compared with short-term interest rates in the United States.

The composition of foreign liquid dollar claims on U.S. residents is related to the functions of foreign dollar balances. Holdings of *demand deposits* in U.S. banks by foreign official institutions and by private nonbanks, which serve mainly the transactions function in both cases, reached a peak in 1968 and thereafter declined, perhaps indicating a tendency to economize on holdings of noninterest-bearing assets in favor of Eurodollars and short-term time deposits in U.S. banks. The very large increase in American liquid dollar holdings of *foreign official institutions* between the end of 1969 and the end of 1971 is accounted for almost entirely by the increase in holdings of U.S. government obligations; holdings of the other categories of liquid dollar assets declined. The vast bulk of the American liquid dollar holdings of foreign *private nonbanks* has taken the form of demand and short-term time deposits in U.S. banks; this suggests that their holdings of American dollar balances mainly perform the functions of transactions and precautionary balances for use in effecting international payments. The nonbanks' other liquid claims, including U.S. government securities, negotiable certificates of deposit (CDs), and other short-term securities, have remained stable at about one billion dollars from 1957 until the end of 1971.

INTRA-MULTINATIONAL BANK LIABILITIES

U.S. liquid liabilities to *foreign commercial banks* require special analysis because of the importance of *intra-multinational bank* liabilities included in the data. Table 2.2 shows that portion of recorded U.S. liquid liabilities to foreign commercial banks constituted by (line 2) the liabilities of U.S. commercial banks to their foreign branches and by (line 4) the liabilities of U.S. agencies, branches and subsidiaries of foreign commercial banks to their head offices and affiliates abroad. Together these liabilities accounted

TABLE 2.2

Analysis of U.S. Liquid Liabilities to Foreign Commercial Banks, 1964–71

(end of period; billions of dollars)

	1964	1965	1966	1967	1968	1969	1970	1971
1. Total liquid liabilities[a]	7.3	7.4	10.0	11.1	14.6	23.8	17.5	11.3
2. Liabilities of U.S. banks to their foreign branches[b]	1.2	1.3	4.0	4.2	5.6	12.6	6.2	1.3
3. Liquid liabilities to foreign commercial banks excluding liabilities to foreign branches of U.S. banks—l. 1 − l. 2	6.1	6.1	6.0	6.9	9.0	11.2	11.3	10.0
4. Liabilities of U.S. agencies, branches and subsidiaries of foreign banking corporations to head offices and affiliates abroad[c]	3.3	3.2	3.5	3.8	4.4	5.6	6.0	4.5
5. Liabilities net of intra-multinational bank liabilities— l. 3 − l. 4	2.8	2.9	2.5	3.1	4.6	5.6	5.3	5.5
6. Demand deposit liabilities of large U.S. commercial banks[d]	n.a.	1.5	1.6	1.8	2.1	2.5	2.4	2.4
7. Residual—l. 5 − l. 6—constituting largely interest-earning liabilities to foreign banks	n.a.	1.4	0.9	1.3	2.5	3.1	2.9	3.1[e]
8. Ratio (in percent) of intra-bank accounting entries to total liquid liabilities— (l. 2 + l. 4)/l. 1	62	61	75	72	68	76	71	51

a. Table 2.1, l. IB4.
b. Table 2.1, l. IB4a.
c. U.S. Department of the Treasury. Includes Western European, Canadian, and Japanese agencies, branches, and subsidiaries.
d. *Federal Reserve Bulletin*, various issues.
e. $2.2 billion constituted interest-earning liabilities and $1.0 billion constituted demand deposits. Data derived from *Federal Reserve Bulletin*, August 1972, p. A80.

for over 70 percent of all recorded U.S. liquid liabilities to foreign commercial banks over the 1966–70 period.

Prior to 1972 these intra-multinational bank liabilities were rather arbitrarily recorded in U.S. statistics. Most were classified as demand deposit liabilities of U.S. commercial banks, some were classified as other types of liabilities, including time deposits. The unsatisfactory classification of these intra-multinational bank liabilities has rendered the recorded U.S. liabilities to *foreign commercial banks* almost meaningless as far as these categories are concerned.

The liabilities of U.S. commercial banks to their foreign branches largely represent borrowings from the Eurodollar market through their foreign branches. These liabilities have fluctuated greatly; they increased from $1.2 billion at the end of 1964 to a high of $14.5 billion at the end of November 1969; thereafter, as U.S. commercial banks repaid their indebtedness to the Eurodollar market, they declined steadily to $1.3 billion at the end of 1971.[7]

The liabilities of U.S. agencies and branches of foreign banks represent funds invested by Canadian and other foreign banks in the U.S. banking business. For the most part these funds do not appear to be highly volatile. During the period from March 1965 to September 1970 they grew very steadily from $3.0 billion to $6.1 billion. During 1971, however, they declined sharply from $6.0 billion at the beginning of the year to $4.5 billion at the end, mainly no doubt as a consequence of the dollar crisis of that year. Nevertheless, the number of U.S. agencies and branches of foreign banks has been increasing recently. According to data published by the Federal Reserve Board, the number of U.S. agencies and branches of foreign banks reporting under the VFCR guidelines rose from 51 at the end of December 1971 to 62 at the end of February 1973. During this same period, the foreign assets of these institutions (held for own account) rose from $3.0 billion to $5.7

---

7. Data on the liabilities of U.S. commercial banks to their foreign branches have been compiled by both the Federal Reserve Board and the U.S. Department of the Treasury from reports made by both the head offices of U.S. commercial banks and by the foreign branches. Estimates from the various series do not agree because of the differences in coverage. We have used the end-of-quarter estimates published in the *Survey of Current Business* for the years 1968–71. Beginning with 1970 these estimates have been based on data gathered by the U.S. Department of the Treasury from head offices of the U.S. banks on the last day of the quarter; prior to that time they were based on Federal Reserve reports of the head offices of U.S. banks. There are two sets of estimates published currently in the *Federal Reserve Bulletin*: (1) end-of-month estimates based on reports from the foreign branches (beginning with the February 1972 issue of the *Federal Reserve Bulletin*); and (2) estimates based on reports by the parent banks. These estimates do not agree even when the reporting date is the same, and neither agrees with the Treasury estimates; the discrepancies reflect differences in reporting methods. See *Federal Reserve Bulletin*, July 1972, pp. A88, A90.

billion, indicating that they have been an important channel for the outflow of U.S. capital.[8]

Beginning with the estimate for December 31, 1971, liabilities of U.S. banks to their foreign branches and liabilities of U.S. agencies and branches of foreign banks to their head offices and affiliates abroad are recorded in the *Federal Reserve Bulletin* as "other short-term liabilities" rather than as (mainly) "deposits" as before. In addition, certain liabilities of the foreign branches to foreign official institutions, formerly included in U.S. liabilities to foreign official institutions, are recorded as U.S. liabilities to foreign commercial banks, also under the heading "other short-term liabilities." [9] This change had the effect of reducing the recorded demand deposits of foreign commercial banks in U.S. commercial banks by about $3.6 billion as of the end of 1971.[10] Unfortunately, data are insufficient to reconstruct the estimates for earlier periods on the revised basis. The data on U.S. liabilities to foreign commercial banks for the period January–September 1971 are also complicated by the issuance of over $3 billion in special U.S. Treasury and Export-Import Bank securities to foreign branches of U.S. banks; these securities were retired in October 1971.[11] We suggest later that for certain purposes it is desirable analytically to treat foreign branches of U.S. banks as a part of the U.S. resident banking system. In the case of the liabilities of U.S. branches of foreign banks to their head offices and affiliates abroad, however, a good portion of these liabilities probably reflects more or less permanent investments in the U.S. banking system.

Despite the confusing aspects of the data arising from intra-multinational bank accounting entries, certain features are apparent. First, exclusive of these entries, foreign commercial bank demand

8. Data taken from a paper by Andrew F. Brimmer entitled "American International Banking: Trends and Prospects," April 2, 1973.

9. Before the reporting system was changed, nearly $800 million of foreign official deposits with foreign branches of U.S. banks (as of December 31, 1971) were reported as U.S. short-term liabilities to foreign official institutions. See *Federal Reserve Bulletin,* August 1972, Table 8, p. A80.

10. The items are recorded both before and after change in coverage for December 31, 1971, in the *Federal Reserve Bulletin,* July 1972, Table 8, p. A80. See also our Table 2.1.

11. These securities were issued in order to reduce the return flow of funds to the Eurodollar market resulting from the repayment of funds borrowed by U.S. banks.

deposits in U.S. banks rose gradually with the growth in international transactions during the 1960s but then apparently declined slightly. This trend is indicated by the growth of foreign commercial bank demand deposits in large U.S. banks as given in line 6 of Table 2.2.[12] After adjustment to exclude intra-multinational bank items, short-term time deposits of foreign commercial banks (included in line IB2 of Table 2.1 and in line 7 of Table 2.2) have probably been less than a billion dollars throughout the 1960s and were only about $0.3 billion at the end of 1971. Holdings of U.S. government securities by foreign commercial banks have been a minor item, but holdings of bankers' acceptances and other short-term securities and negotiable CDs have constituted a substantial portion of foreign commercial bank holdings of U.S. liquid dollar assets.

THE TRANSACTIONS DEMAND FOR FOREIGN DOLLAR BALANCES

Important questions relating to the international role of the dollar arise from the relationship between the growth of foreign dollar balances available for transactions and precautionary purposes on the one hand and the growth of international transactions financed with dollars on the other. We have noted the relative stability of demand and short-term deposits in U.S. banks held by foreign official institutions and of all liquid dollar assets held by private nonbanks over the 1963–71 period. There did occur a modest growth in demand and time deposits of foreign commercial banks in large U.S. banks over most of the period, but these deposits have declined since 1969. Aside from intra-multinational bank dollar liabilities, foreign holdings of U.S. dollar balances related to the transactions and precautionary function have shown only a modest growth compared with the 120 percent rise in the value of U.S. foreign trade between 1963 and 1971, or the rise of 130 percent in the dollar value of world trade, and with the even larger expansion of international financial transactions arising from the growth of capital movements denominated in dollars.

12. Total foreign commercial bank demand deposit holdings in the United States (excluding intra-multinational bank liabilities) were $3.4 billion at the end of 1971 (compared with $2.4 billion held in "large U.S. banks"); no estimates prior to that date are available.

To what extent can it be said that the demand for dollar balances for transactions and precautionary purposes has declined relative to the volume of international dollar transactions? This question has significance for the future demand for dollar balances aside from the demand for dollars as official reserves and for the investment portfolios of foreign private nonbanks. It is possible that the growth of multinational banks and of multinational nonbanking corporations has reduced the demand for the traditional type of foreign dollar balances. Without the expansion of multinational institutions, the growth of foreign deposits in U.S. banks might well have been larger. It also seems likely that Eurodollar deposits have provided a substitute for short-term deposits in the United States.

U.S. Liquid Claims on Foreigners and the U.S. Net Liquidity Position

An analysis of the U.S. liquid claims on foreigners either for the purpose of estimating the U.S. net international liquidity position or for estimating the net American liquid dollar positions of each of the three categories of foreign liquid dollar asset holders presents difficult conceptual problems. While we have adopted with only minor changes the Department of Commerce definition of U.S. liquid liabilities to foreigners, the Department of Commerce classification of U.S. short-term claims on foreigners as between liquid and nonliquid short-term claims is not only arbitrary but inadequate for the purposes we have in mind. For example, foreign-owned bankers' acceptances and negotiable CDs are regarded as U.S. *liquid* liabilities to foreigners, but three-month U.S. bank loans to foreign banks and official institutions and foreign liabilities to the United States arising from (short-term) acceptances made for the account of foreign banks are regarded in U.S. official statistics as short-term *nonliquid* claims on foreigners. Because of the international intermediation role of the United States, short-term liquid liabilities to foreigners tend to exceed short-term liquid claims on foreigners, but just the opposite is the case with short-term nonliquid liabilities and claims as defined by the Department of Commerce.

While recognizing a certain arbitrariness in all definitions, we believe it would be appropriate for the purpose of measuring the net international liquidity position of the United States to include in U.S. liquid claims on foreigners *all* short-term claims on foreign banks and official institutions, whether they take the form of loans or deposits. Therefore, we have added to the Department of Commerce estimates of U.S. short-term liquid claims on foreigners U.S. bank-reported loans to foreign commercial banks and official institutions as well as acceptances (short-term) made for the account of foreigners.[13] According to our definition (detailed in the appendix to this chapter), as of December 31, 1971, U.S. liquid claims on foreigners were $10.8 billion (including U.S. Treasury holdings of convertible currencies); this compares with $4.3 billion in liquid claims on foreigners according to the Department of Commerce definition. (However, U.S. liquid claims on foreign commercial banks are grossly understated since only a fraction of U.S. resident Eurodollars and other foreign currency deposits with foreign banks are recorded.) On this basis the U.S. net liquidity position vis-à-vis foreigners at the end of 1971 was a negative $55.4 billion. Taking into account also the U.S. official holdings of gold, SDRs, and the gold tranche position in the IMF, the U.S. overall international liquidity position was a negative $44.1 billion as contrasted with a positive figure of $11.1 billion at the end of 1957 and a negative position of $2.6 billion at the end of 1963 (see Table 2.1).

The vast bulk of the negative U.S. net international position is accounted for by the net position vis-à-vis *foreign official institutions*—a negative $50.2 billion at the end of 1971. During most of the 1960s the U.S. net liquidity position vis-à-vis foreign official institutions remained fairly constant; it was a negative $13.0 billion at the end of 1969 as contrasted with a negative $14.0 billion at the end of 1963. Moreover, until 1970 this negative position vis-à-vis foreign official institutions was in most years approximately offset by U.S. official gold reserves and the IMF gold tranche position. However, as mentioned previously, the U.S. official reserve transactions account was affected positively by U.S. commercial bank bor-

13. In most cases payments under the acceptances are an obligation of a foreign bank.

rowings from the Eurodollar market, especially in 1966, 1968, and 1969.

The U.S. net negative liquid position vis-à-vis *foreign private nonbanks* remained fairly stable over the 1963–71 period, ranging from $-\$3.2$ billion at the end of 1963 to a high of $-\$4.8$ billion at the end of 1968 and declining to $-\$3.8$ billion at the end of 1971 (see Table 2.1). Although U.S. liquid liabilities to private nonbanks totaled \$4.2 billion at the end of 1971 as against U.S. liquid claims of only \$0.4 billion, total U.S. *short-term* claims on foreign private nonbanks were an estimated \$10.3 billion as against \$7.1 billion in U.S. short-term liabilities (liquid and nonliquid) to foreign private nonbanks at the end of 1971.[14] Therefore the U.S. net short-term position vis-à-vis foreign private nonbanks at the end of 1971 was a positive \$3.2 billion.[15] Moreover, the United States has had a positive short-term position vis-à-vis foreign nonbanks since 1960.

Changes in net liquid liabilities of the United States vis-à-vis *foreign commercial banks* have, of course, been greatly affected by U.S. bank borrowings from their foreign branches. As may be observed by comparing line IB4a with line IIB in Table 2.1, changes in the net liquid position of the United States vis-à-vis foreign commercial banks have rather closely paralleled changes in claims of foreign branches of U.S. banks on their parent banks. Changes in the net liquid position of the United States vis-à-vis foreign commercial banks have also been affected by the liabilities of U.S. branches and agencies of foreign banks to their head offices abroad. Moreover, the sum of the above two categories of intra-multinational bank liabilities has exceeded by a substantial margin the net (negative) U.S. liquid position vis-à-vis foreign commercial banks

---

14. Data for nonliquid U.S. short-term liabilities and claims on foreign private nonbanks are published in the *Federal Reserve Bulletin,* July 1971, Table 14, p. A85; Table 20, p. A87; and Table 26, p. A91. We have assumed that all short-term claims on foreigners in the form of bank-reported "collections outstanding" are liabilities of foreign private nonbanks. Undoubtedly a substantial portion of them represent liabilities of foreign banks, but there is no way of disaggregating the data.

15. The data on U.S. short-term claims do not take into account the large claims of U.S. corporations on their foreign affiliates, many of which should be regarded as short-term claims.

in every year over the period 1963–71, indicating that the United States has had net claims on foreign commercial banks exclusive of intra-multinational bank accounts.[16] As a rule, foreign commercial banks do not take large uncovered positions in foreign currencies. However, a full analysis of the combined American dollar and Eurodollar positions of foreign commercial banks must include an analysis of their Eurodollar operations.

Foreign Holdings of Eurodollars

Prior to the 1960s virtually all foreign holdings of liquid dollar assets took the form of claims on U.S. entities. Although foreign commercial banks in certain countries have for several decades accepted deposits denominated and payable in U.S. dollars, the Eurodollar market as we know it today has been in existence only since 1957 when the London foreign exchange banks began transacting regular business in dollar deposits.[17] Since that time there has been a spectacular rise in dollar deposits held by foreigners with banks outside the United States and in dollar claims by these banks on foreigners. Thus the development of the Eurodollar market has introduced a volume of liquid dollar obligations of foreign commercial banks to other foreign banks and nonbanks which at the end of 1970 was substantially larger than the volume of U.S liquid liabilities to foreigners.[18] It is the purpose of this section to discuss the nature of the Eurodollar market and to present estimates of Eurodollar deposits by category of foreign holder.

16. This is true after taking account of the claims of foreign branches of U.S banks on their head offices and of the claims of foreign banks on their branches and agencies in the United States.

17. See Paul Einzig, *The Eurodollar System*, 3d ed., London: Macmillan, 1967 pp. 2–4.

18. Short-term dollar liabilities (including interbank deposits) to nonresidents excluding U.S. residents, of the commercial banks of eight European countries plus liabilities of Canadian and Japanese banks totaled $64 billion as of the end of 1970 (Bank for International Settlements, *Forty-First Annual Report*, Basle June 1971, pp. 161, 165). To this amount must be added an estimated $9 billion in short-term dollar liabilities of Canadian banks and of the banks in the eight European countries to the residents of the countries in which the banks are located plus other foreign-owned Eurodollar deposits for which data are not available.

NATURE OF THE EURODOLLAR MARKET

Eurodollars are time deposits—denominated and payable in U.S. dollars—held in banks outside the United States, including foreign branches of U.S. banks. Their maturity ranges from "call deposits" to deposits with maturities exceeding one year. The vast bulk of the Eurodollar deposits have a maturity of less than six months. They are not ordinarily transferable through the process of the depositors drawing drafts on their deposits for transfers of funds directly to others. Hence they do not serve as a medium of exchange. Some banks issue Eurodollar certificates of deposit, and there is a substantial secondary market in London for negotiable Eurodollar certificates of deposit.[19]

We may distinguish two levels of the Eurodollar market, each dependent on the other. First, there is the interbank market which is dominated by Western European banks, including the European branches of U.S. banks, and for which London is the principal center. Second, there is the market involving Eurodollar banks on the one hand and nonbank depositors and borrowers on the other.

The interbank market is essentially an interest arbitrage market among commercial banks in which central banks, including the Bank for International Settlements (BIS), participate through swap transactions with commercial banks. Through this market commercial banks adjust their dollar positions over time as required by their obligations arising from nonbank deposits and from their loans to nonbanks and to banks outside the Eurodollar system. The interbank market is somewhat analogous to the Federal Funds market in the United States. It is highly competitive, and traders operate on profit margins of one-eighth of 1 percent or less between the cost and yield on dollar funds.

The system of Eurodollar banks may be regarded as a large financial intermediary. It accepts deposits from individuals, firms, governments, central banks, and commercial banks outside the system, and it makes loans to business firms, government agencies, and commercial banks outside the system. The interest rates paid

19. See E. L. Blacktop, "Sterling and Dollar CDs in London," *Euromoney*, Vol. , October 1969, pp. 24–25.

by Eurodollar banks on these deposits and charged on these loans outside the system tend to fluctuate with the rates established in the interbank market, with allowance for risk and cost factors.

The spread between three-month Eurodollar loan rates to prime borrowers and three-month Eurodollar deposit rates has consistently been under 1 percent. In contrast, in most foreign developed countries, including the European countries, Japan, Australia, and South Africa, the spread has been well over 1 percent and in some countries it is usually in excess of 2 percent.[20] (Canada and the United Kingdom are notable exceptions.) This situation has meant that for most of the countries and in most periods, borrowers have been able to obtain funds at a lower cost and lenders to receive a higher rate of return in the Eurodollar market than in the national money markets.[21] One reason for the smaller spread between the lending and borrowing rates in the Eurodollar market is that Eurodollar banks are not normally required to maintain reserves against Eurodollar deposits. Another factor is that the Eurodollar market is highly competitive on an international basis. Large corporations can borrow at the cheapest rates available throughout the entire Eurodollar market while Eurodollar banks are competing with each other for deposit funds from sources throughout the world. National commercial bank deposit rates are often maintained by collusion among the domestic banks or by regulation; the same may be true of commercial bank loan rates. Finally, the Eurodollar market is a "wholesale" market that exceeds in size that of the domestic money markets of most countries. The greater size and efficiency of the interbank Eurodollar market has narrowed the spread between the cost of funds and the yield on excess funds that Eurodollar banks decide to put into the Eurodollar market.

The Eurodollar banking system may be defined narrowly to include only the commercial banks (including foreign branches of

20. Comparative data on commercial bank deposit rates and lending rates to prime borrowers are given in the monthly issues of Morgan Guaranty Trust Company of New York *World Financial Markets*.

21. Account must be taken of the spread between spot and forward rates on the dollar in terms of the local currency in measuring the differential between national money market rates and covered Eurodollar rates. When the dollar is weak relative to the national currency, borrowers are favored; when the national currency is weak relative to the dollar, lenders to the Eurodollar market are favored.

U.S. banks) of the eight European countries reporting to the BIS, or more broadly to include all banks outside the United States that accept dollar deposits and make dollar loans. However, except for data on the U.S. dollar liabilities and assets of Canadian banks published by the Canadian government and certain data on the assets and liabilities of foreign branches of U.S. banks, our statistical information for the analysis of the operations of the Eurodollar banking system consists mainly of data for commercial banks of the eight European countries reporting to the BIS. These eight European countries are referred to as the "inside area," [22] while the rest of the world is referred to as the "outside area." Although we would like to be able to analyze the operations of Eurodollar banks throughout the world, limitations of data have constrained our analysis mainly to the operations of banks in the "inside area" plus Canada.[23]

A Eurodollar deposit may be held by a foreign individual or private nonbank corporation, by a foreign official institution such as a central bank, or by a foreign commercial bank. Eurodollar deposits are also held by U.S. residents, but in this case they are not "foreign-owned" balances from a U.S. perspective. To the individual holder, a Eurodollar deposit in a foreign branch of a U.S. bank is little different from a liquid claim on the U.S. parent bank. For example, if a foreigner shifts his deposit from the Chase Manhattan Bank in New York to the London branch of the Chase Manhattan

---

22. "Inside area" banks include the commercial banks in Belgium, France, Germany, Italy, the Netherlands, Sweden, Switzerland, and the United Kingdom plus the BIS (the assets of which are included in the data recorded for the Swiss banks).

23. The BIS *Annual Reports* also provide data on the dollar assets and liabilities vis-à-vis nonresidents of the commercial banks of Canada and Japan, but no data are given on the dollar positions of residents of these countries. Data on nonresident dollar liabilities and assets of British banks are regularly reported in the Bank of England *Quarterly Bulletin,* and fairly complete data on the U.S. dollar assets and liabilities of Canadian banks are published by the *Bank of Canada Review* (formerly *Statistical Summary*). Data on dollar liabilities and claims of foreign branches of U.S. banks are gathered and reported by the U.S. Treasury and the Federal Reserve Board. Finally, there are limited data on dollar claims and obligations of commercial banks of other countries usually published in reports of the central banks. Unfortunately there are serious gaps in all of the data series and the data coverage is not consistent between various sources. We have, therefore, limited our statistical analysis of the Eurodollar market to the BIS data, supplemented by certain data from British, Canadian, and U.S. governmental sources and from the IMF.

Bank, he continues to have a dollar claim on the Chase Manhattan Bank even though technically his dollar claim is now on a British resident rather than on a U.S. resident. Eurodollar deposits with foreign commercial banks other than branches of U.S. banks are clearly liabilities of foreign institutions, although to the individual deposit holder it may make little difference whether his Eurodollar deposit is with a British bank or a Swiss bank or a foreign branch of a U.S. bank.

Eurodollar bank loans to foreigners give rise to dollar obligations of one non-U.S. resident to another. Only if the loans are made to U.S. residents do they constitute dollar liabilities of the United States to foreigners. The proceeds of dollar loans by Eurodollar banks may appear initially as a deposit on the books of the lending bank, with payment being made by wire transfer; the dollars can be used for making payments to U.S. residents, to non-U.S. residents, or for increasing the dollar balances of the borrower in a U.S. resident bank or its foreign branch. The borrower frequently converts the dollars into a nondollar currency on the foreign exchange market for financing purchases in his own or a third currency. Such dollars sold on the foreign exchange market may accrue to foreign banks or nonbanks or they may flow into a foreign central bank.

Foreign branches of U.S. banks have played a leading role in the development of the Eurodollar market. The rapid growth in the number and size of these branches has been in response both to the requirements of the parent banks for borrowing funds from the Eurodollar market under conditions of credit stringency in the United States and to the desire to accommodate the banking needs of their corporate customers who have gone abroad. It is unlikely that the Eurodollar market would have achieved anything like its present size and importance in the absence of the expansion of the Eurodollar operations of foreign branches of U.S. banks beginning in the early 1960s. We therefore regard the Eurodollar market in considerable measure as an extension of the U.S. banking system.[24]

24. U.S. dollar liabilities of all foreign branches of U.S. banks (excluding those in the Bahamas) to nonresidents of the United States totaled $31 billion as of December 31, 1971 (including dollar liabilities to residents of the foreign country in which they were domiciled). As of the same date, U.S. dollar liabilities of all "inside area" banks to nonresidents (excluding U.S. residents) were estimated to be $65 billion. (BIS 1972 *Annual Report,* p. 151) Although not all foreign

The Eurodollar market constitutes the dominant part of the larger Eurocurrency market. Nondollar Eurocurrencies, including bank deposits denominated in Deutsche marks, Swiss francs, sterling, and gulden, among others, are time deposits in banks domiciled in countries other than the country whose currency is involved. Between December 31, 1969, and December 31, 1971, nonresident Eurocurrency liabilities of the European banks reporting to the BIS rose from $56.8 billion to $97.9 billion, and the nondollar portion rose from 19 percent to 28 percent over the same period. Part of the increase in the nondollar portion reflected the currency revaluation in 1971, but even disregarding the effects of the currency revaluation, the relative importance of *Eurodollars* declined. Despite the external weakness of the dollar during 1970 and 1971, the Eurodollar market, as measured by the BIS, has continued to grow although at a slower rate than in 1968 and 1969.

THE BIS CONCEPTUAL FRAMEWORK OF
THE EURODOLLAR MARKET

The conceptual framework formulated by the staff of the BIS for estimating the "net size" of the Eurodollar market is based on the dollar liabilities (sources of dollars) and dollar assets (uses of dollars) of the commercial banks in the eight European countries reporting to the BIS, that is, the "inside area" banks as a group.[25] The framework is illustrated by the organization of data for the 1964–71 period, given in Table 2.3. Sources of dollar funds to the "inside area" banks are divided into two general categories, namely, "outside area" sources and "inside area" sources. (This conceptual framework is obscured by the geographical classification of sources in Table 2.3.) The "outside area" sources consist of deposits, or other short-term claims,[26] held by commercial banks, central banks,

branches of U.S. banks are located within the "inside area," the "inside area" branches of U.S. banks account for about 85 percent of all liabilities of foreign branches of U.S. banks (excluding the Bahama branches).

25. BIS, *Thirty-Ninth Annual Report,* Basle, June 1969, Chap. V.

26. In calculating both U.S. sources and uses, the BIS has adjusted the data on U.S. short-term claims on and liabilities to European reporting banks so as to exclude working and compensating balances with U.S. banks and loans and credits from the United States to the "inside area" banks regarded as not "related to the growth of the Eurodollar market." *Ibid.*

26   *Foreign Dollar Balances: A Review*

TABLE 2.3

Estimated Net Size of the Eurodollar Market
in Eight European Countries, 1964–71
(end of period; billions of dollars)

	Dec. 1964	Dec. 1965	Dec. 1966	Dec. 1967	June 1968	Dec. 1968	June 1969	Dec. 1969	Dec. 1970[a]	Dec. 1971[b]
Sources										
United States	0.6	0.7	1.1	1.7	2.9	3.2	4.4	3.8	4.2	5.7
Canada	0.9	0.6	0.6	0.9	1.0	1.3	2.3	2.9	3.8	4.0
Subtotal	1.5	1.3	1.7	2.6	3.9	4.5	6.7	6.7	8.0	9.7
Western Europe										
"inside area"	4.4	6.6	8.4	9.6	12.2	13.2	16.8	18.3	21.0	23.4
Nonbanks	1.8	2.2	2.8	4.0	4.8	5.2	8.1	9.8	9.7	10.8
Banks[c]	2.6	4.4	5.6	5.6	7.4	8.0	8.7	8.5	11.3	12.6
Other Western Europe	n.a.	n.a.	1.1	1.4	1.5	1.9	2.1	2.7	4.5	4.9
Subtotal	n.a.	n.a.	9.5	11.0	13.7	15.1	18.9	21.0	25.5	28.3
Japan	—	—	—	0.1	0.1	0.1	0.2	0.4	0.6	0.9
Eastern Europe	0.3	0.3	0.4	0.4	0.4	0.6	0.6	1.0	1.0	1.2
Other	n.a.	n.a.	2.9	3.4	4.4	4.7	7.1	8.4	10.9	13.4
Subtotal	n.a.	n.a.	3.3	3.9	4.9	5.4	7.9	9.8	12.5	15.5
Total	9.0	11.5	14.5	17.5	22.5	25.0	33.5	37.5	46.0	53.5
Uses										
United States	1.8	2.0	4.4	5.2	8.8	9.5	16.7	16.5	12.7	8.0
Canada	0.4	0.7	0.6	0.7	0.9	0.9	1.2	1.3	2.3	1.8
Subtotal	1.2	2.7	5.0	5.9	9.7	10.4	17.9	17.8	15.0	9.8
Western Europe										
"inside area"	5.0	6.3	6.3	6.9	7.1	7.9	8.9	11.6	17.4	21.9
Nonbanks	2.3	3.3	3.7	4.1	4.5	4.7	5.1	5.6	10.1	12.5
Banks[d]	2.7	3.0	2.6	2.8	2.6	3.2	3.8	6.0	7.3	9.4
Other Western Europe	n.a.	n.a.	0.9	1.2	1.4	1.5	1.3	1.6	2.6	4.0
Subtotal	n.a.	n.a.	7.2	8.1	8.5	9.4	10.2	13.3	20.0	25.9
Japan	0.4	0.5	0.6	1.0	1.4	1.7	1.4	1.5	2.3	3.1
Eastern Europe	0.5	0.5	0.7	0.8	0.8	0.9	0.9	1.0	1.7	2.4
Other	n.a.	n.a.	1.0	1.7	2.1	2.6	3.1	3.9	7.0	12.3
Subtotal	n.a.	n.a.	2.3	3.5	4.3	5.2	5.4	6.4	11.0	17.8
Total	9.0	11.5	14.5	17.5	22.5	25.0	33.5	37.5	46.0	53.5

SOURCES: BIS *Annual Reports* for 1969 (*Thirty-Ninth*), 1970 (*Fortieth*), 1971 (*Forty-First*), and 1972 (*Forty-Second*).

a. Some of the data for December 1970 have been derived indirectly from the BIS *Forty-First Annual Report*, 1971.

b. The BIS *Forty-Second Annual Report*, 1972, did not break down the table on the estimated size of the Eurocurrency market by Eurodollars and other Eurocurrencies. It gave only the approximate net size of the Eurodollar market as $54 billion as of December 31, 1971. The breakdown in this column has been derived and partly estimated from other data given in the BIS *Forty-Second Annual Report*. In estimating "inside area" nonbank sources and uses, it was assumed that the ratio of dollar "inside area"

Footnotes from Table 2.3 continued

nonbank sources and uses to total Eurocurrency "inside area" nonbank sources and uses was the same as the ratio of all dollar nonbank sources and uses to total Eurocurrency nonbank sources and uses, respectively. This may not have been the case so that, for example, dollar "inside area" nonbank sources may be overstated while dollar "inside area" nonbank uses may be understated.

c. Deposits by official monetary institutions (including the BIS) of the reporting area and conversions by the banks of domestic or third-currency funds into dollars.

d. Conversions by the banks of Eurodollar funds into domestic or third currencies.

and private nonbanks outside the eight European reporting countries with the commercial banks in the reporting area ("inside area" banks). The "inside area" sources consist of deposits of private nonbanks within the reporting area (both residents and nonresidents of the countries of the commercial banks holding the deposits) and "bank" sources.[27] For this purpose, the "bank" sources include deposits of the BIS, deposits of "inside area" central banks, and dollars injected into the market by the "inside area" banks themselves. The latter are derived by the conversion of domestic or third currencies into dollars by the "inside area" banks, in part through swap transactions with central banks. "Inside area" interbank deposits are excluded from both the sources and uses of dollars and therefore from the BIS calculation of the "net size of the market," which is defined as equal to the sum of the sources or the sum of the uses.[28] The reasons for eliminating "inside area" interbank deposits are similar to those advanced for deducting interbank deposits in computing the domestic money supply. It should be stressed that not all of the dollar sources of the market are nonbank Eurodollar deposits. The sources generated by the "inside area" banks by the conversion of nondollar currencies into dollars and lent directly to customers or passed through to other "inside area" banks for lending to customers do not originate with nonbank deposits. However, we do not know the exact volume of these sources since a portion of the "bank" sources given in Table 2.3 constitute deposits of "inside area" central banks and the BIS.

The BIS concept of uses parallels its concept of sources. "Out-

27. The BIS terminology becomes confusing and, at this point, misleading, in view of what is included in and excluded from "bank" sources, as further explained in the text.

28. For an analysis of the BIS concept of the "net size of the market," see BIS *Annual Report* for 1969 (*Thirty-Ninth*), pp. 147–151.

side area" uses consist of dollar claims of "inside area" banks on both banks and nonbanks outside the reporting area, while "inside area" uses are divided between claims on nonbanks and "bank" uses. The latter item (designated by "banks" under the heading "inside area" uses in Table 2.3) represents the amount of dollars obtained from various sources which are converted by the banks into domestic or third currencies for loans to customers. Hence, not all of these uses generate dollar obligations to "inside area" banks, just as not all of the sources constitute dollar deposit liabilities of "inside area" banks. These anomalies in the BIS concept of the "net size of the market" raise complications in the calculation of the volume of foreign dollar liquidity in the form of Eurodollar deposits.

In its 1971 annual report, the BIS for the first time provided data on liabilities and assets of "inside area" banks vis-à-vis all nonresidents,[29] banks and nonbanks. The item for banks lumps together both nonresident commercial bank and official institution assets and liabilities vis-à-vis "inside area" banks,[30] but fortunately there are estimates of Eurodollar holdings of official institutions in other sources. The BIS also publishes data on the assets and liabilities of "inside area" banks vis-à-vis all nonbanks within the eight reporting countries. U.S. government data on assets and liabilities of foreign branches of U.S. banks provide data which differentiate among foreign commercial banks, nonbanks, and foreign official institutions, but not on a geographical basis. Canadian banking statistics provide a breakdown between banks and nonbanks on a geographical basis. Unfortunately, therefore, it is impossible to combine the data from these various partial sources so as to provide a complete breakdown of all Eurodollar bank liabilities and assets vis-à-vis commercial banks, official institutions, and nonbanks, and on a geographical basis.

The dollar and nondollar components of the Eurocurrency market are in effect all part of the same market, and the BIS concept of the net size of the Eurodollar market in terms of sources and

---

29. That is, residents other than of the country in which the reporting bank is located. Unless further specified, "nonresidents" in the BIS terminology may reside either in another "inside area" country or outside the area.

30. *Forty-First Annual Report,* p. 159.

uses of dollars, as distinct, say, from the sources and uses of Euro-Deutsche marks, is something of an artificial abstraction. For example, if the "source" of dollars to the market is a Deutsche mark deposit converted into dollars by a Eurocurrency bank, the Deutsche mark deposit is regarded as a "source" in the calculation of the net sizes of both the Eurodollar market and the Euro-Deutsche mark market. Conversely, if a dollar deposit with a Eurocurrency bank is converted into Deutsche marks and lent to a customer, the Deutsche mark loan becomes both a "use" of dollars and a "use" of Deutsche marks. This system of accounting may lead to a misinterpretation of the table published in the BIS 1971 annual report (p. 164) showing the estimated size of both the Eurodollar market and of the total Eurocurrency market. Thus, it would not be correct to subtract the items included in the estimated size of the Eurodollar market in order to obtain a corresponding value for the combined nondollar Eurocurrency market.

For the reason stated above, the BIS 1972 annual report (p. 155) did not disaggregate the Eurocurrency market by giving separate estimates, as in previous annual reports, for dollar and nondollar "sources" and "uses." It did, however, give an overall estimate of the "net size" of the Eurodollar market alone for the end of 1971. Thus the data given in Table 2.3 for the "net size" of the market at the end of 1971 were derived by the authors from other tables on dollar liabilities and assets given in the BIS 1972 annual report. Since our concern is with Eurodollar liabilities and assets and not with other Eurocurrencies, the above consideration does not affect the validity of the BIS concept of the Eurodollar market.[31]

VOLUME AND COMPOSITION OF FOREIGN LIQUID DOLLAR ASSETS REPRESENTED BY EURODOLLAR DEPOSITS

Not all Eurodollar deposits should be included in the concept of foreign liquid dollar assets any more than all domestic deposits in commercial banks should be included in the definition of money. In determining what Eurodollar deposits should be included we are faced with many of the same problems that arise in measuring the volume of domestic money. The definition of foreign liquid dollar

31. The above analysis is based on private correspondence with a member of the staff of the BIS.

assets should be functionally oriented, and Eurodollar deposits perform several international monetary functions. Although not ordinarily employed for financing international trade and other transactions, Eurodollars are highly liquid and are denominated in the principal international currency. Consequently they serve to economize on the use of U.S. demand deposits required for transactions purposes and are a good substitute for other highly liquid American dollar assets, e.g., time deposits, required as precautionary balances. Eurodollar deposits of longer maturity provide investors with a liquid dollar asset which is a close substitute for American liquid dollar assets of similar maturity but they yield a higher return with little additional risk or conversion cost. Eurodollars also provide central banks with an international reserve asset which may yield a higher return or other advantages over American liquid dollar assets. Finally, Eurodollar deposits are a substitute for holding liquid assets denominated in other currencies. Much the same functions could be performed by deposits and loans denominated in any international currency. In fact, there are some economists who would question our inclusion of Eurodollars in an analysis of foreign liquid dollar assets on the grounds that Eurodollars are nothing more than bank deposits in foreign banks that happen to be denominated in dollars as a convenient numéraire.[32] We reject this position for three reasons. First, Eurodollar deposits constitute a specific claim on American dollars, and the proceeds of Eurodollar loans are quickly transferred by the borrowers for making dollar payments or for conversion into other currencies rather than held as working balances. Second, Eurodollar deposits are dollar liabilities of banks and therefore affect the behavior of banks regarding their assets. Third, the Eurodollar market is closely tied to the U.S. monetary system, largely as a consequence of the dominant role of foreign branches of U.S. banks in the market.

Foreign nonbank concerns and individuals, official institutions,

---

32. For example, Ira Scott states that ". . . the so-called Euro-dollar market is neither European nor a market for dollars. It is, rather, the market for bank deposits which are denominated in foreign currencies." (Ira O. Scott, *The Eurodollar Market and Its Public Policy Implications,* Joint Economic Committee, Congress of the United States, 91st Congress, Second Session, USGPO, Washington, D.C., February 25, 1970, p. 2.)

and commercial banks may all utilize Eurodollar deposits for one or more of the functions of international money described above. Nevertheless, in calculating the volume of Eurodollars that effectively serve these functions it is necessary to avoid certain duplications. As has been noted, the BIS considers the commercial banks in the eight reporting countries ("inside area" banks) as constituting a system of financial intermediaries and therefore excludes their claims on each other from its calculation of the "net size of the market." Two objections may be made to this approach. First, the system of financial intermediaries represented by the banks in the "inside area" is arbitrarily defined since banks in other Western European countries, Canada, the Bahamas, and in certain other areas of the world participate in one degree or another in the international financial intermediary system in which Eurodollar funds deposited in one bank are shifted by redepositing in other banks so as to equate the demand with the supply of Eurodollar funds. Logically, therefore, the BIS concept of the "net size of the market" ought to be expanded to include all of the banks that accept dollar deposits and play an active role in the interbank market. Such an expansion would require both the inclusion of additional Eurodollar deposits held by these banks in the enlarged system and the deduction of additional interbank deposits in the calculation of the BIS concept of the "net size of the market." [33] However, the definition of "inside area" banks has been dictated in considerable measure by the availability of data to the BIS—data on the Eurocurrency operations of commercial banks in the eight countries that are regularly reported to that institution.

33. The BIS does not publish an estimate of "inside area" interbank deposits employed in the calculation of the "net size of the market." However, we may estimate such interbank deposits to be approximately $20 billion as of December 31, 1969. This estimate is found by deducting from the $25.6 billion in liabilities of "inside area" banks to "inside area" nonresidents an estimated $4.9 billion in liabilities of "inside area" banks to "inside area" nonresident nonbanks. By way of clarification, " 'inside area' nonresidents" are nonresidents of the country in which the individual reporting bank is located; it does not mean nonresidents of the entire "inside area." (The $25.6 billion figure is given by the BIS *Fortieth Annual Report,* p. 156. The $4.9 billion estimate is based on estimates of "inside area" bank liabilities to "inside area" nonresident nonbanks for the years December 1964–December 1967 given in earlier BIS Annual Reports, but not given for 1968 and 1969 in the *Fortieth Annual Report.*)

The second problem in the BIS concept, at least for the purposes of the present analysis, arises from the fact that the Eurodollar liabilities of foreign branches of U.S. banks are actually dollar liabilities of U.S. banks, even though they are technically not liabilities of U.S. residents. In one sense, foreign branches of U.S. banks are part of the U.S. banking system; in another sense, they operate as nonresident banks. The resources of a foreign branch are legally available to the parent bank, and the parent bank is responsible for the liabilities of its branches. Yet there are constraints on transfers of funds between a branch and its parent arising out of both the Voluntary Foreign Credit Restraint Program (VFCR) and the special reserve requirements on transfers of funds from foreign branches to parent banks established by the Federal Reserve Board in 1969. Except during periods when their parent banks are borrowing heavily from the Eurodollar market, most of the deposits of foreign branches of U.S. banks are employed for making loans to non-U.S. residents. Nevertheless, dollar deposits in foreign branches of U.S. banks have at times been transferred to the parent for use in the domestic money market, and foreign branches often make loans that otherwise would have been made by the parent bank. In a variety of ways foreign branches of U.S. banks serve as channels whereby U.S. monetary conditions are transferred abroad. This is also true of agencies and branches of foreign banks operating in the United States.

Foreign branches of U.S. banks play a dual role in the Eurodollar market. On the one hand, their Eurodollar deposit liabilities are the liabilities of U.S. banks to the depositors and are virtually equivalent to American liquid dollar assets. On the other hand, the foreign branches are a part of the system of Eurodollar banks that carry on a large volume of redepositing with one another in the process of shifting Eurodollar funds from nonbank lenders to the ultimate borrowers. These interbank deposits arising out of the financial intermediary functions of the system of Eurodollar banks should not be included in the calculation of foreign dollar liquidity. Nevertheless, foreign commercial banks hold dollar deposits with foreign branches of U.S. banks as precautionary balances or for other purposes as a substitute for holding balances with the parent banks in the United States. Therefore, we believe that a portion of the foreign commercial bank holdings of Eurodollar deposits in foreign branches of U.S. banks should be regarded as foreign dollar liquidity on a par with foreign commercial bank balances held with U.S. resident banks. Perhaps the best measure of this element is the *net* deposit liabilities of foreign branches of U.S. banks to foreign commercial banks. These net deposit liabilities rose from about $0.8 billion at the end of 1965 to $9.7 billion at the end of 1969 but fell to $4.0 billion at the end of 1971 (see Table 2.4). Changes in net

TABLE 2.4

Foreign Branches of U.S. Banks: Dollar Liabilities
to Foreign Commercial Banks and Dollar Deposits
in Foreign Banks, 1965–71*
(end of year; millions of dollars)

	Liabilities to Foreign Commercial Banks		Deposits in Foreign Commercial Banks		Net	
	All Branches	U.K. Branches	All Branches	U.K. Branches	All Branches	U.K. Branches
1965	1,845	1,536	1,028	843	817	693
1966	3,419	2,762	1,210	999	2,209	1,763
1967	4,409	3,700	1,815	1,506	2,594	2,194
1968	7,553	6,121	2,241	1,838	5,311	4,283
1969	17,793	13,302	8,074	6,265[a]	9,719	7,037[a]
1970	19,807	13,684	13,248	9,420[a]	6,559	4,264[a]
1971	22,069	14,038	18,040	12,762[a]	4,029	1,276[a]

SOURCES: For 1965–68, *Treasury Bulletin*, November 1970; for 1969, 1970, and 1971, *Federal Reserve Bulletin*, February 1972, pp. 110–113, and July 1972, pp. A88–A89.
* Data for 1965–68 and 1969 and 1970 are not exactly comparable.
a. Includes deposits of foreign official institutions.

liabilities of the foreign branches to foreign commercial banks have closely paralleled changes in net claims of the branches on U.S. residents, including parents. Details on these relationships will be given in Chapter 3.

Turning now to the *foreign nonofficial nonbank* Eurodollar deposits, clearly all such deposits should be included in our concept of foreign liquid dollar assets. Unfortunately, data on such deposits are mainly limited to those held with "inside area" and Canadian banks.[34] We estimate that Eurodollar deposits with these banks held by foreign nonofficial nonbanks throughout the world rose from $7.0 billion at the end of 1966 to $17.6 billion at the end of 1970, but then fell to $14.5 billion at the end of 1971 (see Table 2.5 and explanatory note). During the expansion phase the bulk of the growth in these deposits took place in 1969, when these deposits grew by $7.9 billion. Eurodollar deposits of *foreign private non-banks* rose proportionately more during 1969 than any other source of Eurodollar funds contributing to the growth of the market. Moreover, foreign private nonbank deposits held by the residents of the "inside area" countries with "inside area" banks accounted for $4.6 billion (or about 60 percent) of the $7.9 billion increase in foreign private nonbank Eurodollar deposits during 1969. Another $1.7 billion was accounted for by the increase in (non-U.S. resident) private nonbank Eurodollar deposits in Canadian banks, leaving only $1.6 billion to be accounted for by private nonbank depositors in the rest of the world outside the United States. Apparently "inside area" residents were attracted by the high interest rates on Euro-dollar deposits relative to those available in the domestic markets, while the risk of holding dollars seemed small at a time when the U.S. official reserve transactions balance was in substantial surplus and "inside area" central banks were losing dollars. During 1970 "inside area" private nonbanks reduced their Eurodollar deposits slightly, and the growth of funds available for Eurodollar lending to non-U.S. residents was provided mainly by the repayment of dollars to the market by U.S. banks and by Eurodollar deposits made by foreign official agencies.

In estimating private nonbank Eurodollar deposits, we must recognize that the ultimate holders of deposits are frequently not known to the reporting banks, and that the figures given in Table 2.5 are probably understated.[35] Moreover, there are dollar deposits by nonofficial nonbanks held in commercial banks in areas of the

---

34. We also have data on total foreign nonbank Eurodollar deposits with foreign branches of U.S. banks. Although the bulk of these deposits are *included* in the BIS data on "inside area" banks, we have no satisfactory way of determining the amount *excluded* from the BIS data.

35. Individuals and nonbanking concerns frequently hold Eurodollar deposits through banks acting as trustees so that the "inside area" and Canadian banks may report such deposits as being held by foreign banks.

TABLE 2.5

Estimates of Nonofficial Nonbank Eurodollar Deposits
With "Inside Area" and Canadian Banks, Excluding U.S.
Resident Deposits, 1966–71
(end of period; billions of dollars)

	1966	1967	1968	1969	1970	1971
1. Inside area bank liabilities to inside area nonbanks[a]	2.8	4.0	5.2	9.8	9.7	10.8
2. Inside area bank liabilities to all nonresident nonbanks[b]	4.1	4.7	6.2	10.5	11.2	10.0
3. Inside area bank liabilities to inside area nonresident nonbanks[c]	1.4	2.0	2.6	4.9	4.8	5.4
4. Inside area bank liabilities to U.S. resident nonbanks[d]	0.5	0.8	2.3	2.7	3.1	4.0
5. Total inside area bank liabilities to nonbanks, excl. U.S. res. (l. 1 + l. 2) − (l. 3 + l. 4)	5.0	5.9	6.5	12.7	13.0	11.4
6. Canadian bank U.S. dollar liabilities to nonbanks excluding U.S. residents[e]	2.0	2.4	2.8	4.5	4.6	3.1
7. Total (l. 5 + l. 6)	7.0	8.3	9.3	17.2	17.6	14.5

SOURCES: BIS *Annual Reports* and *Bank of Canada Review*, various issues.

NOTE: In the construction of Table 2.5, we began with the BIS figures for "inside area" bank liabilities to "inside area" nonbanks (line 1) and for "inside area" bank liabilities to all nonresident nonbanks (line 2). Line 2 includes that portion of line 1 which constitutes "inside area" bank liabilities to "inside area" nonresident nonbanks (line 3), an amount which had to be estimated. Hence in order to avoid double counting we subtracted line 3 from the sum of lines 1 and 2. In addition, in order to exclude U.S. resident deposits we also subtracted line 4. In this way we obtained total "inside area" bank liabilities to all nonbanks, excluding U.S. resident nonbanks (line 5). To this figure we added Canadian bank U.S. dollar liabilities to nonbanks, excluding U.S. residents (line 6), and obtained estimates of Eurodollar deposits held by foreign firms and individuals with "inside area" banks and Canadian banks (line 7).

"Inside area" bank liabilities to U.S. resident nonbanks (line 4) are very rough estimates of that portion of U.S. sources in Table 2.3 which constitutes U.S. nonbank sources. There are statistics on the liabilities of foreign branches of U.S. banks to U.S. residents other than their parent banks; these include liabilities to both U.S. banks and nonbanks and, in any case, probably include no more than half of the U.S. nonbank dollar deposits with "inside area" banks recorded in the BIS data. See *Federal Reserve Bulletin*, current issues, and *Treasury Bulletin*, November 1970. (Actually the BIS data greatly underestimate U.S. resident sources since a large portion of the U.S. Eurodollar deposits are hidden in trustee accounts. However, the important thing for our estimate of foreign nonbank Eurodollar deposits is to subtract from the BIS estimate of total nonbank deposits that amount which represents U.S. nonbank deposits included in the BIS data.)

There are also statistics on U.S. short-term liabilities to each of the eight "inside area" countries reported by U.S. banks (*Federal Reserve Bulletin*, various issues). We have reason to believe that these liabilities consist largely of liabilities of U.S. banks or other liabilities not recorded as nonbank Eurodollar deposits. The esti-

36 *Foreign Dollar Balances: A Review*

Footnotes from Table 2.5 continued

mates in line 4 are derived by taking the larger of: (*a*) liabilities of foreign branches of U.S. banks to U.S. residents other than their parent banks; or (*b*) the BIS figure for U.S. sources (Table 2.3) *less* U.S. short-term liabilities to the eight "inside area" countries reported by U.S. banks.

    a. Data from Table 2.3.

    b. Data for all years from BIS, *Forty-Second Annual Report*, 1972, p. 151.

    c. Estimates are based on the assumption that one-half of "inside area" bank liabilities to "inside area" nonbanks are to "inside area" nonresident nonbanks. This was roughly the proportion for 1966 and 1967. (See BIS, *Thirty-Eighth Annual Report*, June 1968, p. 154.)

    d. Years 1968–71 estimated by subtracting U.S. short-term claims on "inside area" countries reported by U.S. banks from BIS estimates of "inside area" bank dollar liabilities to U.S. residents. Years 1966 and 1967 are dollar liabilities of foreign branches of U.S. banks to U.S. residents other than parent banks.

    e. Calculated from *Bank of Canada Review*, various issues.

world other than the "inside area" countries and Canada, but we have little information as to their amount. Such deposits may amount to a billion dollars in the Bahamas[36] and to half a billion dollars or more in Asian cities such as Singapore, Hong Kong, and Tokyo.[37] There are also Eurodollar holdings in banks in Beirut, which is relatively free of exchange controls. In most of the remainder of the world, however, foreign exchange controls prohibit individuals and nonbanking concerns from holding foreign currencies beyond the amounts needed for actual trading purposes, and most of these foreign currencies tend to be held directly in the countries whose currencies are involved. Domestic banks are usually not permitted to maintain dollar accounts for residents, and nonresidents are more likely to maintain their Eurodollar deposits with European banks.

A third category of Eurodollar deposits which should be included in our concept of foreign liquid dollar assets is that of *foreign official* Eurodollar holdings. Such holdings are usually included in official reserves along with their holdings of American liquid dollar assets, sterling assets, and liquid assets denominated in other convertible

---

    36. Data on foreign nonbank deposits (in all currencies combined) in Bahama branches of U.S. banks are given in the *Federal Reserve Bulletin,* current issues.

    37. The center of the Asian dollar market is Singapore, where banks are estimated to hold between $300 and $400 million in dollar deposits. See S. A. Pandit, "The Asian Dollar and Free Gold Markets in Singapore," *Finance and Development,* No. 2, 1971, pp. 32–36.

currencies.[38] Foreign central banks and other foreign official institutions tend to hold their Eurodollar or other Eurocurrency deposits with the commercial banks of other countries or with the BIS, which in turn may deposit these funds in European commercial banks. Since the data on dollar assets and liabilities of Swiss banks (as reported in the BIS *Annual Reports*) include those of the BIS, there is no direct way of knowing the volume of central bank dollar funds that are channeled into the Eurodollar market through the BIS. However, there is indirect evidence that the BIS increased its Eurocurrency deposits with commercial banks by about $2 billion during 1970.[39] Since the BIS reduced its American liquid dollar holdings by approximately $1 billion during 1970, these dollars may have constituted one source of the funds that went into the Eurocurrency market.[40] The BIS *Forty-First Annual Report* estimates that total official placements in the Eurocurrency market increased by about $7 billion during 1970; the bulk of this increase was in Eurodollars.[41]

In its *1972 Annual Report* the IMF published estimates of aggregate Eurodollar holdings of foreign official institutions for the years 1964–71. The estimated official Eurodollar claims shown in Table 2.6 are based on the IMF information from 58 countries. Identified official Eurodollar claims rose from $1.3 billion at the end of 1964 to $10.1 billion at the end of 1971.[42] Gross liquid

38. The foreign exchange component of official international reserves is published by country in the monthly issues of IMF, *International Financial Statistics*, Washington, D.C. In most cases the official foreign exchange holdings include central bank holdings of Eurocurrencies, but such holdings are not separately identified.

39. This estimate is based partly on the fact that BIS time deposits with other banks and advances (mainly to commercial banks) rose by approximately $2 billion during the fiscal year ended March 31, 1971. *Forty-First Annual Report*, 1971, p. 183.

40. BIS holdings of short-term assets in the United States may be estimated roughly from the table in the *Federal Reserve Bulletin* titled "Short-term Liabilities to Foreigners Reported by Banks in the United States, by Country" under the item "Other Western Europe." See *Federal Reserve Bulletin*, June 1971, p. A75.

41. *Forty-First Annual Report*, 1971, p. 166.

42. Unidentified official Eurocurrency holdings and residual sources of reserves totaled $8.7 billion at the end of 1971, some of which undoubtedly represented official Eurodollar deposits. IMF, *1972 Annual Report*, Washington, D.C., 1972, p. 30.

38   *Foreign Dollar Balances: A Review*

TABLE 2.6

Liquid Dollar Positions of Foreign Official Institutions, 1964–71
(end of period; billions of dollars)

	Estimated Eurodollar Claims	Claims on U.S. Residents	Liabilities to U.S. Residents	Net Position vis-à-vis U.S. Residents $(2)-(3)$	Gross Liquid Dollar Claims $(1)+(2)$	Net Combined Position $(1)+(4)$
	(1)	(2)	(3)	(4)	(5)	(6)
1964	1.3	15.8	0.7	15.1	17.1	16.4
1965	1.4	15.8	1.1	14.7	17.2	16.1
1966	2.0	14.9	1.6	13.3	16.9	15.3
1967	2.4	18.2	2.7	15.5	20.6	17.9
1968	3.6	17.3	3.8	13.5	20.9	17.1
1969	4.3	16.0	3.0	13.0	20.3	17.3
1970	9.2	23.8	0.7	23.1	33.0	32.3
1971	10.1	50.7	0.5	50.2	60.8	60.3

SOURCES: For Eurodollar claims, see IMF, *1972 Annual Report*, Washington, D.C., 1972, Table 10, p. 30. Data on liquid claims and liabilities vis-à-vis the United States from Table 2.1 of this study.

dollar claims of foreign official institutions increased from $17.1 billion at the end of 1964 to $60.8 billion at the end of 1971; the bulk of this increase took place during 1970 and 1971. So long as foreign commercial banks and private nonbanks do not increase their demand for American liquid dollar assets, dollars placed in the Eurodollar market by foreign official institutions quickly return to the central banks, with a consequent increase in the combined American dollar and Eurodollar holdings of foreign central banks. However, the process of redepositing dollars in the Eurodollar market by foreign official institutions and the utilization of the Eurodollar funds for loans throughout the world may have some effect on the distribution of American liquid dollar asset holdings among foreign countries. The nature of this redistribution of foreign official holdings of American liquid dollar assets is difficult to determine because of the large speculative movements that occurred during 1970 and 1971.

We have so far identified three categories of Eurodollar deposits which we believe should be included in any measure of foreign liquid dollar assets, namely: (1) net foreign commercial bank Eurodollar positions with foreign branches of U.S. banks; (2) gross foreign nonofficial nonbank holdings of Eurodollars; and (3) gross foreign official holdings of Eurodollars.[43] There is a fourth category of Eurodollar deposits that should be considered as a part of foreign liquid dollar holdings, namely, Eurodollar deposits held by foreign commercial banks outside the Eurodollar banking system with foreign commercial banks within the system (other than the already covered foreign branches of U.S. banks). The analysis of this fourth category of liquid dollar holdings involves difficult conceptual problems and virtually insoluble data problems. One solution would be to regard all commercial banks in the world outside the United States as a part of the Eurodollar banking system and exclude from our concept of foreign liquid dollar assets all foreign interbank deposits with the exception of the net Eurodollar position of foreign commercial banks with foreign branches of U.S. banks. However, it would not be realistic to include in the system commercial banks that are outside the Eurodollar interbank redepositing system. Banks outside this system borrow Eurodollar funds either for their own account or on behalf of their customers in much the same way that a nonbank borrows. Yet any geographical delineation of the Eurodollar interbank system is likely to be just as arbitrary as the BIS definition which limits the interbank system to the commercial banks in the eight European reporting countries. Moreover, if we designate a geographical area as representing the Eurodollar banking system, say, Western Europe, the Bahamas, Canada, and Japan, we are unable to make a reasonably reliable estimate of the amount of Eurodollar deposits of foreign commercial banks outside the area so designated with commercial banks within that area. Thus, just as we must omit from our calculations those nonbank Eurodollar deposits with foreign commercial banks outside the European reporting area and Canada, so must we omit a substantial volume of Eurodollar deposits held by foreign commercial banks

43. Unlike U.S. official data on foreign official American dollar holdings, neither IMF nor BIS data on Eurodollar holdings of foreign official institutions include the Eurodollar holdings of the BIS (see p. 9, footnote 6).

outside the Eurodollar interbank system with commercial banks within the system (other than foreign branches of U.S. banks which have already been accounted for).[44]

NET EURODOLLAR POSITIONS

Table 2.7 shows the growth of Eurodollar borrowings of foreign nonbanks from "inside area" and Canadian banks from $5.8 billion at the end of 1966 to $24.8 billion at the end of 1971. By subtracting foreign nonbank Eurodollar deposits with "inside area" and Canadian banks we obtained the net Eurodollar position of foreign nonbanks. This position was in approximate balance in 1968 but rose to a positive balance of $6.3 billion at the end of 1969. Between the end of 1969 and the end of 1971 foreign nonbank Eurodollar deposits declined while foreign nonbank Eurodollar borrowings more than doubled, so that by the end of 1971 foreign nonbanks had a negative net Eurodollar position of $10.3 billion.

Table 2.8 presents estimates of the net Eurodollar positions of foreign commercial banks (excluding foreign branches of U.S. banks), with foreign branches of U.S. banks and with foreign nonbanks and their Eurodollar liabilities to foreign official agencies.[45]

44. We can suggest an order of magnitude for Eurodollar deposits of foreign commercial banks outside of Western Europe, Canada, and Japan with "inside area" banks. (Eurodollar deposits of foreign commercial banks outside Western Europe, Canada, and Japan are probably very largely with "inside area" banks.) The BIS has estimated the dollar liabilities of "inside area" banks to depositors outside of Western Europe, Canada, Japan, and the United States to be $11.9 billion as of the end of 1970 (see Table 2.3), but there are no data on how these deposits are allocated among those held by official institutions, by commercial banks, or by nonbanks; nor do we know what proportion of these deposits is with U.S. bank branches located in the "inside area." We have reason to believe, however, that at least $4 billion represents liabilities to official institutions. Our analysis of the volume of "outside area" nonbank Eurodollar deposits held with "inside area" banks in the estimates given in Table 2.5 suggests that about $3 billion of the nonbank Eurodollar deposits with "inside area" banks is held by nonbanks in the areas outside of Western Europe, Canada, Japan, and the United States. This leaves about $5 billion for Eurodollar deposits of foreign commercial banks located in the countries outside of Western Europe, Canada, Japan, and the United States which is held with "inside area" banks, of which perhaps 40 percent is held with U.S. foreign branches. Thus we arrive at an estimate of $3 billion in Eurodollar deposits at the end of 1970 for our fourth category of foreign liquid dollar assets. However, this could vary by a billion dollars either way. No estimate for this category of Eurodollar deposits is given in our tables.

45. No data are available on foreign commercial bank loans to foreign official institutions, but they are not believed to be large.

TABLE 2.7

Estimates of Foreign Nonofficial Nonbank Eurodollar
Liabilities to "Inside Area" and Canadian Banks,
and of Net Eurodollar Position, 1966–71
(end of period; billions of dollars)

	1966	1967	1968	1969	1970	1971
1. Inside area bank loans to inside area nonbanks[a]	3.7	4.1	4.7	5.6	10.1	12.5
2. Inside area bank loans to all nonresident nonbanks[b]	2.1	3.4	5.2	6.1	11.9	14.4
3. Inside area bank loans to inside area nonresident nonbanks[e]	0.8	0.7	0.9	1.1	2.0	2.5
4. Inside area bank loans to U.S. resident nonbanks[d]	0.4	0.5	1.0	1.1	1.5	1.4
5. Total inside area bank loans to nonbanks, excl. U.S. residents (l. 1 + l. 2) − (l. 3 + l. 4)	4.6	6.3	8.0	9.5	18.5	23.0
6. Canadian bank U.S. dollar loans to nonbanks, excl. U.S. residents[e]	1.2	1.1	1.2	1.4	1.7	1.8
7. Total (l. 5 + l. 6)	5.8	7.4	9.2	10.9	20.2	24.8
8. Net Eurodollar position of nonbanks[f]	1.2	0.9	0.1	6.3	−2.6	−10.3

SOURCES: BIS *Annual Reports*; *Federal Reserve Bulletin*; and *Bank of Canada Review*, various issues.

a. Data from Table 2.3, "inside area" nonbank uses.

b. Data for all years from BIS, *Forty-Second Annual Report*, 1972, p. 151.

c. For 1966 and 1967, estimates are taken from BIS, *Thirty-Eighth Annual Report*, June 1968, p. 154. Estimates for years 1968, 1969, 1970 and 1971 are based on the assumption that 20 percent of "inside area" bank loans to "inside area" nonbanks are to "inside area" nonresident nonbanks. This was roughly the proportion for 1966 and 1967 combined.

d. Very rough estimates were determined by using the lesser of (*a*) short-term liabilities to the eight "inside area" countries reported by U.S. nonbanking concerns and (*b*) claims of foreign branches of U.S. banks (excluding Bahama branches) on U.S. residents other than parents. While both of these series include items other than "inside area" bank loans to U.S. nonbank residents and exclude some "inside area" bank loans to U.S. nonbank residents, the two series tend to move together at least during the 1968–71 period. In any case, the estimates are probably understated.

e. Calculated from *Bank of Canada Review*, various issues.

f. Table 2.5, line 7, less Table 2.7, line 7.

These estimates are only rough approximations owing to the limitations of the data. The net dollar positions with foreign nonbanks apply only to "inside area" and Canadian banks, excluding the (roughly estimated) positions of foreign branches of U.S. banks. The net dollar position of foreign commercial banks with foreign

TABLE 2.8

Net Eurodollar Positions of Foreign Commercial Banks,
Excluding Foreign Branches of U.S. Banks, 1966–71*
(end of period; billions of dollars)

	1966	1967	1968	1969	1970	1971
1. Net position with foreign branches of U.S. banks[a]	2.2	2.6	5.3	9.7	6.6	4.0
2. Net dollar position of "inside area" and Canadian banks vis-à-vis all foreign nonbanks, excl. position of foreign branches of U.S. banks[b]	−1.0	−0.8	−0.7	−4.6	0.7	5.8
3. Eurodollar liabilities to foreign official agencies, excl. liabilities of foreign branches of U.S. banks[c]	−0.9	−1.3	−2.3	−2.7	−5.5	−5.1
Memorandum items:						
1. Net dollar position of foreign branches of U.S. banks with foreign nonbanks[d]	−0.2	−0.1	0.6	−1.7	1.9	4.5
2. Liabilities of foreign branches of U.S. banks to foreign official agencies[d]	1.1	1.1	1.3	1.6	3.7	5.0

* Data are not available for determining overall net Eurodollar position.

a. See Table 2.4, net position of all foreign commercial banks.

b. Table 2.7, line 8, with sign reversed, adjusted for the net position of foreign branches of U.S. banks vis-à-vis foreign nonbanks. (The data assume the net position of all foreign branches of U.S. banks vis-à-vis foreign nonbanks is equal to the net position of foreign branches in the "inside area" alone.)

c. From Table 2.6, adjusted for liabilities of foreign branches of U.S. banks to foreign official agencies.

d. Data for 1966–68 from *Treasury Bulletin*, November 1970, pp. 124–137; data for 1969, 1970, and 1971 from *Federal Reserve Bulletin*, February 1972 and July 1972. It should be noted that the two series are not fully comparable.

branches of U.S. banks declined sharply during 1970 and 1971 while their liabilities to foreign official institutions rose. On the other hand, the net Eurodollar positions of "inside area" and Canadian banks (excluding the positions of foreign branches of U.S. banks) vis-à-vis foreign nonbanks shifted from negative $4.6 billion at the end of 1969 to positive $5.8 billion at the end of 1971. This shift came about mainly as a consequence of the sharp rise in loans to foreign nonbanks relative to foreign nonbank Eurodollar deposits. The increase in loans to foreign nonbanks was heavily concentrated in loans to residents of the countries in which the "inside area" banks are located (Table 2.7, line 1 minus line 3). Since the increase in resident deposits of "inside area" banks was much smaller (Table 2.5, line 1 minus line 3), the "inside area" banks developed a substantial surplus ($4.6 billion) with their own resident nonbanks.

The net Eurodollar positions of *foreign commercial banks* (excluding foreign branches of U.S. banks) with the three categories given in Table 2.8 are not fully comparable since the net positions with foreign branches of U.S. banks and with foreign official agencies include all foreign commercial banks, while the net position with foreign nonbanks includes only "inside area" and Canadian banks. Moreover, we have no way of estimating the net position of "inside area" banks (excluding foreign branches of U.S. banks) with "outside area" commercial banks not regarded as members of the Eurodollar interbank system.[46] Therefore, lines 1, 2, and 3 in Table 2.8 cannot properly be added to obtain a net overall Eurodollar position of foreign commercial banks.

Foreign Holdings of American Dollars
and Eurodollars Combined

In Table 2.9 we have combined, by three categories of ownership, foreign liquid dollar assets represented by Eurodollars with the estimates of foreign holdings of American liquid dollar assets detailed in Tables 2.1 and 2.2. During the period from the end of 1966 to the end of 1971, total foreign liquid dollar assets, as we have defined them, rose by $57.1 billion. Of this increase, $17.4 billion was accounted for by Eurodollar deposits. Of the latter, $8.1 billion represented the increase in Eurodollar deposits of foreign official institutions; $1.8 billion the increase in Eurodollar claims of foreign commercial banks on foreign branches of U.S. banks; and $7.5 billion the increase in Eurodollar deposits represented by nonofficial nonbanks. (As has been noted, however, our estimates of Eurodollar deposits of nonbanks are limited to those with "inside area" and Canadian banks.)

46. Interbank deposits among members of the Eurodollar redepositing system would be excluded from our concept of the volume of foreign dollar liquidity just as interbank deposits among "inside area" banks are excluded in the BIS concept of the "net size of the market."

TABLE 2.9

Estimates of Foreign Liquid Dollar Assets by Category of Asset Holder and by Type of Assets, 1966–71
(end of period; billions of dollars)

	1966	1967	1968	1969	1970	1971
Foreign Official Institutions						
Claims on U.S. residents	14.9	18.2	17.3	16.0	23.8	50.7
Eurodollar claims	2.0	2.4	3.6	4.3	9.2	10.1
Subtotal	16.9	20.6	20.9	20.3	33.0	60.8
Foreign Commercial Banks						
U.S. liquid dollar assets, excl. foreign branches of U.S. banks[a]	6.0	6.9	9.0	11.2	11.3	10.0
Net position with foreign branches of U.S. banks	2.2	2.6	5.3	9.7	6.6	4.0
Subtotal	8.2	9.5	14.3	20.9	17.9	14.0
Foreign Nonbanks						
Claims on U.S. residents	4.3	4.7	5.0	4.6	4.7	4.2
Eurodollar deposits with "inside area" and Canadian banks[b]	7.0	8.3	9.3	17.2	17.6	14.5
Subtotal	11.3	13.0	14.3	21.8	22.3	18.7
Total	36.4	43.1	49.5	63.0	73.2	93.5
U.S. liquid dollars	25.2	29.8	31.3	31.8	39.8	64.9
Eurodollars	11.2	13.3	18.2	31.2	33.4	28.6

SOURCES: Tables 2.1, 2.2, 2.4, 2.5, and 2.6. *Federal Reserve Bulletin*, February 1972, pp. 110–111, and July 1972, p. A88, and *Treasury Bulletin*, November 1970, pp. 127ff., for assets and liabilities of foreign branches of U.S. banks.
a. Table 2.2, line 3.
b. Includes Eurodollar deposits with foreign branches of U.S. banks.

During the period under consideration, the largest increase in foreign liquid dollar assets ($43.9 billion), including both American dollars and Eurodollars, accrued to foreign official institutions, but over 90 percent of this increase took place during 1970 and 1971. Liquid dollar assets of foreign nonbanks rose by $7.4 billion from the end of 1966 to the end of 1971; all of this increase was accounted for by Eurodollar deposits. Foreign commercial banks increased their total liquid dollar holdings over the period by $5.8 billion, $4.0 billion of which represented an increase in American liquid dollar assets. Of the latter amount, $1.0 billion was accounted for by the increase in liabilities of U.S. agencies and branches of foreign banking corporations to their head offices and branches abroad (see Table 2.2).

Combined foreign holdings of liquid dollar assets show a fairly steady increase over the 1966–71 period, but the components as between Eurodollars and American liquid dollar assets and as among the three categories of foreign liquid dollar asset holders have behaved erratically over the period (Table 2.9). Thus, during 1969, foreign holdings of American liquid dollar assets remained almost stationary, but foreign holdings of Eurodollars nearly doubled. In 1970, Eurodollar holdings increased only modestly with all of the increase accounted for by the increase in foreign central bank holdings. During this same year there was a sharp increase in American liquid dollar holdings of foreign official institutions. Holdings of foreign commercial banks and foreign nonbanks were virtually unchanged. In 1971, foreign holdings of American liquid dollar assets soared as a consequence of the increase in the holdings of foreign official institutions, but foreign holdings of Eurodollars declined. The reader may be puzzled by the fact that foreign holdings of Eurodollars showed a decline between the end of 1969 and the end of 1971 while the BIS estimate of the "net size of the market" rose during that period. Prior to 1970, the two estimates, while different in absolute amounts, tended to follow a similar trend. The reasons for the divergence in the two series are to be found in the differences in both concept and statistical measurement. These differences are explained in Appendix B at the end of this chapter.

We lack sufficient data for determining the net position of all foreign liquid dollar holders combined, but the analytical significance of such an estimate is doubtful. However, in the light of the speculative movements against the dollar in 1970 and 1971, it is of some interest to note the shift of foreign nonbanks from a large net positive position in American dollars and Eurodollars combined ($10.6 billion) at the end of 1969 to a substantial net negative position ($6.5 billion) at the end of 1971 (see Table 2.1, line IIIC, and Table 2.7, line 8). A substantial portion of the increased Eurodollar borrowings by foreign nonbanks during 1971 was reportedly undertaken for the purpose of acquiring strong nondollar currencies for speculative purposes.

As has been noted, our data are inadequate to determine the net Eurodollar position of foreign commercial banks according to the concept we have used. However, it seems unlikely that most foreign commercial banks had large uncovered positive dollar positions in 1970 and 1971. Of the three categories of foreign liquid dollar asset holders, only foreign official institutions had large net (uncovered) positive positions, and these institutions absorbed the vast bulk of the loss in terms of foreign currencies arising from the effective depreciation of the dollar in 1971.

Appendix A: The Conceptual Basis for the Classification of Liquid Claims

FOREIGN LIQUID CLAIMS ON U.S. RESIDENTS

In the accompanying table, foreign liquid claims on U.S. residents are identical with the items included in foreign liquid claims by the Department of Commerce[47] except for two items, namely, nonmarketable, nonconvertible U.S. Treasury bonds and notes, and long-term liabilities to foreigners reported by U.S. banks. Both of these liability items to foreign official agencies, while referred to as nonliquid liabilities, are included in the calculation of the U.S. official reserve transactions balance and hence are below-the-line items in the U.S. net liquidity balance. However, the same liability items to foreign commercial banks and nonbanks are regarded as nonliquid in the Department of Commerce balance of payments format. This treatment appears to us to be inconsistent. Therefore, we have regarded both of these items as foreign liquid claims whether the liabilities are to foreign official agencies, or to foreign commercial banks, or to nonbanking entities. Most foreigners regard long-term time deposits and certificates of deposit of U.S. banks as liquid. In any case, they are readily convertible into cash. Nonmarketable, nonconvertible U.S. Treasury bonds and notes are held entirely by foreign official institutions except for $153 million (as of December 31, 1972) held by German commercial banks.

47. For Department of Commerce definitions see its "Explanatory Notes for Tables 2 and 3," *Survey of Current Business,* June 1951, pp. 51ff.

U.S. LIQUID CLAIMS ON FOREIGNERS

There are just two significant differences from the Department of Commerce concept of liquid claims in our classification of U.S. liquid claims on foreigners. First, we have included U.S. short-term bank loans to foreign official agencies and foreign commercial banks. Second, we have included U.S. bank-reported (short-term) acceptances made for the account of foreigners. It is our view that all U.S. short-term claims on foreign banks and foreign official agencies should be regarded as liquid claims. There are undoubtedly a number of other short-term claims on foreign commercial banks that have not been reflected in our tables, but there is no way of separating them from claims on nonbanks which we have regarded as nonliquid. For example, a substantial portion of U.S. bank-reported "collections outstanding" represents claims on foreign banks. In addition, the volume of U.S. nonbank claims on foreign commercial banks is grossly understated in U.S. official data.

CLASSIFICATION

I. Foreign Claims on U.S. Residents[48]
  A. Foreign central banks and other official institutions
    1. Demand deposits
    2. Time deposits with maturity of one year or less
    3. U.S. government obligations
      a. U.S. Treasury bills and certificates
      b. Marketable U.S. government bonds and notes
      c. Nonmarketable, convertible (into Treasury bills) U.S. Treasury bonds and notes
      d. Nonmarketable, nonconvertible U.S. Treasury bonds and notes[49]

48. "Foreigners" exclude international and regional organizations of which the United States is a member. Liquid claims may be denominated in either dollars or foreign currencies.

49. With the exception of small amounts held by German commercial banks, all nonconvertible U.S. Treasury bonds and notes are held by foreign official institutions.

These claims, and those listed in lines IA5, IIA2, IIB3, and IIB4, below, are defined as nonliquid in *Survey of Current Business* balance-of-payments tables.

4. Other short-term liabilities, including bankers' acceptances, commercial paper, and negotiable certificates of deposit
5. Time deposits and certificates of deposit with a maturity of one year or more

B. Foreign commercial banks, including foreign branches of U.S. banks (same as in A above)

C. Other foreigners (same as in A above except that nonbank concerns and individuals do not own nonmarketable U.S. Treasury bonds and notes)

II. U.S. Liquid Claims on Foreigners

A. Claims on foreign central banks and other official institutions
1. U.S. official holdings of foreign convertible currencies
2. Short-term loans to foreign official institutions reported by U.S. banks
3. Short-term foreign government securities[50]

B. U.S. claims on foreign commercial banks including foreign branches of U.S. banks
1. Short-term deposits with foreign banks reported by U.S. banks
2. Short-term deposits with foreign banks reported by U.S. nonbanking concerns
3. Short-term loans to foreign banks reported by U.S. banks
4. Acceptances (short-term) made for account of foreigners reported by U.S. banks
5. Other short-term claims on foreign banks reported by U.S. banks (and included in Department of Commerce definition of U.S. liquid claims on foreigners)

C. U.S. claims on other foreigners
1. Short-term commercial and finance paper representing obligations of foreign nonbanks reported by U.S. banks[51]

---

50. Available data do not permit disaggregation of U.S. holdings of foreign short-term securities as between foreign government securities and others. Hence in our tables all U.S. resident holdings of foreign short-term securities are listed as obligations of foreign nonbanks.

51. See note 50.

2. Negotiable and other readily transferrable foreign obligations payable on demand or having a contractual maturity of not more than one year from the date on which the obligation was incurred by the foreigner

Appendix B: Relation Between Eurodollar Balances in Table 2.9 and the BIS "Net Size of the Market"

THE reader may be puzzled by the fact that while the BIS "net size of the market" (Table 2.3) increased by $39.0 billion (or by $34.4 billion if U.S. sources are excluded) between the end of 1966 and the end of 1971, our own estimates show that foreign Eurodollar deposits rose by only $17.4 billion over the same period (Table 2.9). This difference has arisen since 1969. Until then the two series did not differ substantially in absolute value, as shown by the following table (in billions of dollars):

	1966	1967	1968	1969	1970	1971
BIS "net size of the market" adjusted to exclude U.S. sources	13.4	15.8	21.8	33.7	41.8	47.8
Foreign Eurodollar balances in Table 2.9	11.2	13.3	18.2	31.2	33.4	28.6

During 1970 and 1971, however, the BIS estimate rose by $14.1 billion, while foreign Eurodollar balances under our concept declined by $2.6 billion. The reasons for this contrasting behavior are to be sought in differences both in the conceptual framework behind the two sets of estimates and in the data coverage. These differences may be further explored in the light of the more detailed comparisons given in Table 2.10.

As for the conceptual aspects, it will be recalled that the BIS "sources" measure all Eurodollar deposits with "inside area" banks less interbank deposits among "inside area" banks, plus those sources of dollar funds generated by "inside area" banks themselves, i.e., not originating from deposits. Our own concept regards the Eurodollar system as encompassing the entire world, i.e., all of the commercial banks that comprise the interbank redepositing sys-

TABLE 2.10

Foreign Holdings of Eurodollars on Alternative Definitions
of Member Banks Comprising the System
(end of year; billions of dollars)

Holder of Eurodollar Deposits and Where Held	"Inside Area" Basis as Defined by BIS[a]			Worldwide Basis Attempted in Present Study		
	1966	1969	1971	1966	1969	1971
I. Claims of foreign nonbanks						
A. On "inside area" banks (incl. foreign branches of U.S. banks)						
1. By residents of the area	2.8	9.8	10.8	2.8	9.8	10.8
2. By nonresidents	2.2	2.9	.6	2.2	2.9	.6
B. On Canadian banks	—	—	—	2.0	4.5	3.1
C. On other "outside area" Eurodollar banks	—	—	—	n.a.	n.a.	n.a.
Subtotal	5.0	12.7	11.4	7.0	17.2	14.5
II. Claims of foreign commercial banks, both inside and outside system, on foreign branches of U.S. banks (net)	—	—	—	2.2	9.7	4.0
III. Other claims of foreign commercial banks outside system						
A. On "inside area" banks				n.a.	n.a.	n.a.
B. On Canadian banks	—	—	—	n.a.	n.a.	n.a.
C. On other "outside area" Eurodollar banks	2.8[b]	12.5[b]	23.8[b]			
IV. Claims of "outside area" central banks	—	—	—	n.a.	n.a.	n.a.
V. Claims of "inside area" central banks and of BIS				} 2.0[c]	4.3[c]	10.1[c]
VI. Eurodollar banks' purchases of dollars	5.6	8.5	12.6	—	—	—
Total	13.4	33.7	47.8	11.2	31.2	28.6

SOURCES: Line IA1 from Table 2.5, line 1. Line IA2 from Table 2.5, line 5 minus line 1. Line IB from Table 2.5, line 6. Line II from Table 2.8, line 1. Lines III and IV combined from Table 2.3, sum of entries for Canada, Other Western Europe, Japan, Eastern Europe, and Other. Lines IV and V combined from Table 2.6, column 1. Lines V and VI combined from Table 2.3, entry for "banks."

a. Includes Belgium, France, Germany, Italy, the Netherlands, Switzerland, Sweden, and the United Kingdom.

b. Includes only IIIA and IV.

c. Excludes claims of BIS.

m.[52] In our approach, all interbank deposits in the worldwide urodollar system are excluded in the calculation of Eurodollar oldings except for net deposits of Eurodollar banks with foreign ranches of U.S. banks. These net dollar liabilities are called Euro-ollars, but they are also net dollar liabilities of the U.S. banking stem. Net liabilities of foreign branches of U.S. banks to foreign ommercial banks include net liabilities to foreign commercial banks utside the Eurodollar banking system as well as those inside. Ac-ording to our concept, the dollar deposits of foreign commercial unks outside the system with all Eurodollar banks inside the system rould also be included in the measure of Eurodollar deposits. How-er, we have no data on the Eurodollar deposits of foreign com-ercial banks outside the Eurodollar system with foreign commer-al banks inside the system, so we have had to omit such deposits om our series.

Using both BIS and Canadian data, we have estimated Euro-ollar deposits of foreign nonbanks with "inside area" and Canadian unks. Admittedly, this estimate omits foreign nonbank deposits in her areas, e.g., Hong Kong, Singapore, the Bahamas, and Beirut, it the volume is probably under $2 billion. Most countries out-le of Western Europe do not permit their domestic banks to ac-pt dollar deposits from residents and most nonresident Eurodollar posits are with "inside area" banks. Our data on Eurodollar de-osits of central banks are from IMF sources; there are no separate IS data on such deposits.[53]

We can now provide at least a partial indication of why the two ries diverge so sharply after 1969. Unfortunately, for reasons ven in explaining the derivation of the tables in Chapter 2, the mponent series cannot be further disaggregated, at least in any stematic fashion, so as to permit a closer comparison.

---

52. A term broader in its geographic implications than *Euro*dollars would be sirable when applied to a worldwide system. Fritz Machulp has employed the m *Xeno-currencies* for bank deposits denominated in currencies other than t of the country in which the bank is located. See Fritz Machlup, "The Euro-llar System and Its Control," *International Monetary Problems,* Washington, C.: American Enterprise Institute for Public Policy Research, 1972, Part I.

53. In the BIS data, Eurodollar deposits of "inside area" central banks and the S are included in "bank" sources, while "outside area" central bank deposits included in "outside area" sources. The IMF data do not include Eurodollar posits of the BIS. (See notes to Table 2.3)

Starting at the top of Table 2.10 with item IB, the Canadian bank figures on nonbank deposits, included in keeping with our world wide approach, account for a decline of $1.4 billion in our serie in 1970 and 1971. As already noted, we should have liked to in clude item IC, but we lack the necessary data, nonbank Eurodolla deposits in still other "outside area" banks.

The next major item (II) included in our estimates but not it those of the BIS—net claims of foreign commercial banks on for eign branches of U.S. banks—shows a decline of $5.7 billion it 1971 and 1972. The logic of including these claims in statistics or foreign holdings of Eurodollars (though not necessarily in the "ne size of the market") has already been given in Chapter 2. It ha also been noted that this decline was a direct consequence of th reduction in U.S. bank borrowings from their foreign branches.

Item III of Table 2.10 shows an omission that may be of som consequence in our estimates, reflecting our inability to derive from the available sources figures on Eurodollar claims of "outside area commercial banks on banks within the Eurodollar system as w conceive it. There is no basis for judging how much these claims if they could be included, would increase our estimates or affec their movement in recent years. We may only note that the omis sion is of a dual nature: (1) we have no data on the claims of "out side area" commercial banks on Eurodollar banks brought within the system under our more global concept; and (2) we cannot em ploy the aggregative data given by the BIS on the basis of its re porting banks, since (apart from other difficulties noted in the nex paragraph) such aggregates would include Eurodollar claims c banks that we think should now be regarded as part of the system.

The converse of the latter point is that the BIS reporting system if it includes claims by banks that could themselves now be con sidered as insiders, risks overstating both the size and the rate c growth of the market. An appraisal of this possibility is, howeve seriously handicapped because the BIS reports do not distinguis between claims by outside area commercial banks and outside are central banks. Consequently, the two are lumped together in Tabl 2.10, showing an increase of $11.3 billion in 1970 and 1971. Fro the data published by the IMF it appears that of the total increas in Eurodollar deposits by foreign central banks in 1970 and 197 (also included in our estimates) about $5 billion came from "outside area" central banks.[54] That would leave roughly $6 billion coming from "outside area" commercial banks out of the BIS estimates in Table 2.10.

The BIS does not give separate figures for Eurodollar deposits of "inside area" central banks. It gives only a series which includes "inside area" central bank and BIS Eurodollar deposits with "inside area" banks combined with the Eurodollar banks' purchases of dollars (including swap deals with central banks). The combined series—identified by the BIS simply as "bank" sources (Table 2.3) —shows an increase of $4.1 billion in 1970 and 1971. We may estimate, however, that of this amount, $3.5 billion consisted of dollar funds generated by the banks themselves—and not included in our estimates—as distinguished from "inside area" central bank deposits.[55]

To summarize, the major sources of difference in the behavior of the BIS estimates of the "net size of the market" and our own estimates of foreign Eurodollar balances between the end of 1969 and the end of 1971 are (*a*) the differential treatment of foreign commercial bank deposits; (*b*) the inclusion in the BIS concept of dollar funds obtained from sources other than Eurodollar deposits; and (*c*) the inclusion in our estimates of foreign nonbank Eurodollar deposits in Canadian banks.

54. IMF, *Annual Report* for 1972, Table 10, p. 30. Central banks of primary producing countries accounted for an increase of $3.7 billion in Eurodollar deposits over the two-year period, and those of industrial countries other than the Group of Ten for an increase of $0.4 billion. In addition, the Bank of Japan may have increased its Eurodollar deposits by some $0.5 billion on the assumption that it accounted for most if not all of the increase in Eurodollar deposits from Japanese sources shown by Table 2.3 between the end of 1969 and the end of 1971.

55. According to the IMF, *op. cit.,* p. 30, Eurodollar holdings of central banks of the Group of Ten countries rose by $1.1 billion over the two-year period. We may assume, as noted above, that some $0.5 billion of this increase came from the Bank of Japan. On this basis, "inside area" central banks accounted for only $0.6 billion of the $4.1 billion from "bank" sources. Conceivably a portion of the remaining $3.5 billion represented an increase in BIS Eurodollar deposits with "inside area" Eurobanks, but these deposits are also not included in our own estimates of central bank deposits derived from IMF sources.

3

The Behavior
of Foreign Dollar Holdings

Introduction

OUR analysis of foreign holdings of liquid dollar assets reveals the complex composition of the types of liquid dollar assets and the categories of debtors and creditors having liquid dollar positions. Changes in the volume and composition of foreign liquid dollar holdings can be explained only in terms of the functional relationships among the various categories of transactors involved in these creditor-debtor relationships. Figure 3.1 shows the network of dollar claims between U.S. and foreign residents and among foreign residents confronting the analyst concerned with the demand and supply of foreign liquid dollar balances. Each of the categories and subcategories of foreign liquid dollar holders has a unique asset preference function, and there are special relationships between each of the pairs of categories of liquid dollar creditors and debtors. There are also special markets for particular types of liquid dollar assets, each operating within a unique institutional framework. In addition to the limitations imposed by the inadequacy of the data, our knowledge of these functions, relationships, and institutions is too fragmentary to permit construction of a complete model of the foreign demand and supply of liquid dollar assets. The most we can hope to do is formulate partial models confined to limited sets of relationships among the variables involved.

In this chapter we are concerned with the origins of foreign dollar holdings, including both American dollar and Eurodollar deposits, and with the relationship of changes in these holdings to the U.S. balance of payments. Closely related to the origins and composi-

54

tion of foreign dollar holdings is the substitutability among foreign holdings of American dollars, Eurodollars, and other currencies. We shall present empirical evidence relating to the degree of substitutability. Also considered here is the influence of the Eurodollar market on the U.S. balance of payments. Finally, we deal briefly with the impact of U.S. and foreign monetary policies and controls on the Eurodollar market.

FIGURE 3.1

Network of Dollar Claims

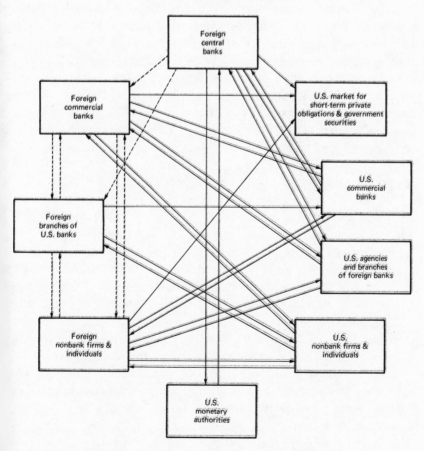

Arrows point from dollar claimant to debtor. Dotted lines are dollar claims between foreign residents.

Origins of Foreign Dollar Holdings

U.S. BALANCE OF PAYMENTS AND FOREIGN HOLDINGS
OF U.S. DOLLARS

Total foreign holdings of American dollars are, of course, always a function of the U.S. balance of payments as a whole. However, a distinction should be made between changes in U.S. liquid liabilities to foreigners which are the result of autonomous transactions in the U.S. international accounts and those that are accommodating or causally determined by the sum of all the other transactions. Changes in foreign nonofficial American dollar holdings are (except possibly for very brief periods) always autonomous, while in recent years a large proportion of the American dollar holdings of foreign official institutions have been accommodating in the sense that, given the existing pattern of exchange rates and the existing foreign nonofficial demand for American dollars, foreign official institutions are the residual buyers and sellers of American dollars. Prior to August 15, 1971, the central banks of the major countries were constrained from converting their dollars into gold at the U.S. Treasury by the knowledge that too great a U.S. gold loss would lead the United States to close the gold window. After August 15, 1971, foreign central banks as a group had no choice but to hold whatever dollars were generated by U.S. balance-of-payments deficits, again given the existing pattern of exchange rates. Individual foreign official institutions often sought to reduce their American dollar holdings by redepositing the dollars in the Eurodollar market. However, no amount of foreign central bank depositing in the Eurodollar market could reduce their collective holdings of American dollars, unless such deposits in some way affected either the foreign nonofficial demand for American dollars or U.S. resident borrowing from the Eurodollar market. Otherwise, the dollars loaned by Eurodollar banks from such deposits would tend to be sold for foreign currencies and thus to accrue once more to the reserves of central banks. The result of such depositing would be, therefore, to increase the sum of American dollars and Eurodollar holdings combined in foreign official reserves and to add to the plethora of dollars.

The foreign nonofficial demand for American liquid dollar assets is determined in part by the transactions and precautionary demand for dollars and in part by the investment demand for American liquid dollar assets. This latter demand is a function of the total foreign demand for liquid assets and of such factors as relative rates of return, risk, and legal constraints which determine the composition of liquid asset portfolios. Under present conditions, a reduction in the foreign nonofficial demand for liquid American dollar assets will not reduce total foreign holdings of American dollars; the excess supply of American dollars is simply absorbed by the foreign central banks. This, of course, assumes that other transactions in the U.S. balance of payments are not affected so as to reduce the supply of American dollars in the hands of foreigners.

THE GENERATION OF EURODOLLAR DEPOSITS

While changes in total foreign holdings of American dollars must always be related in some way to the U.S. balance of payments, changes in the volume of Eurodollar deposits held by foreigners need have no relationship to the U.S. balance of payments. Nor does the volume of Eurodollar deposits have any necessary relationship to the volume of American dollars held by foreign banks. A depositor can acquire Eurodollar deposits with any convertible currency. If he does not have American dollars, he can readily acquire them in the foreign exchange market, or the Eurobank receiving the deposit will denominate it in dollars in exchange for whatever currency is tendered. In either instance, will not the purchase in the foreign exchange market of American dollars by either the depositor or by the Eurobank receiving the deposit increase foreign nonofficial holdings of American dollars? Only momentarily, since the American dollars are in most cases quickly lent and then sold by the borrower in the exchange market for other currencies. There are, of course, cases where the borrower decides to increase his holdings of American dollars. In such cases there will be a net increase in the foreign nonofficial demand for American dollars. However, there is nothing in the expansion of Eurodollar deposits and loans per se that will necessarily give rise to a net increase in the foreign nonofficial demand for American dollars. Hence, additional foreign holdings of American dollars need not be generated by the U.S. balance of payments to support the growth of the Euro-dollar market. This is not to deny, however, that changes in the U.S. balance of payments may affect the Eurodollar market, and activities in the Eurodollar market may have an impact on the U.S. balance of payments. More will be said shortly on these relation-ships.

It has been argued that the maximum potential growth of Euro-dollar deposits is determined by the volume of American dollars available to the Eurobanks, since the Eurobanks require dollar re-serves against Eurodollar deposits. According to this view, the maximum volume of Eurodollar deposits that can be created is some multiple of the volume of American dollar holdings of the Eurodollar banks, analogous to the maximum expansion of com-mercial bank deposits in a domestic monetary system on the basis of a given volume of commercial bank reserves.[1] Those who reject this view argue that the proportion of Eurodollar loans redeposited in the Eurodollar market is relatively small, and that for most countries there are no legal reserve requirements on Eurodollar deposits as there usually are on domestic currency deposits. More-over, Eurodollar banks need not maintain precautionary reserves against Eurodollar deposits in U.S. banks; they can just as well keep dollar reserves in the form of deposits with foreign branches of U.S. banks (which themselves are Eurodollar banks). In fact, since 1969 foreign commercial banks have had larger deposits with for-eign branches of U.S. banks than holdings of American liquid dollar assets. Foreign branches of U.S. banks require no special reserves against dollar deposits since they can draw on the resources of their parent banks for meeting their deposit obligations. Thus, to a de-gree, the ultimate reserves against Eurodollar deposits are the reserve assets of the U.S. banking system.

Our approach to the problem of the limits of the generation of

1. For an exposition of this view, see Milton Friedman, "The Eurodollar Mar-ket: Some First Principles," *The Morgan Guaranty Survey,* October 1969, pp. 4ff. For a contrary view, see Fred H. Klopstock, "Money Creation in the Euro-dollar Market—A Note on Professor Friedman's Views," *Monthly Review,* Vol. 52, Federal Reserve Bank of New York, January 1970, pp. 12–15. For an ex-cellent summary of the various arguments relating to Eurodollar creation, see Fritz Machlup, "The Eurodollar System and Its Control," *Papers and Proceed-ings of a Conference on International Monetary Problems,* Washington, D.C.: American Enterprise Institute for Public Policy Research, 1972.

Eurodollar deposits is that Eurocurrency deposits constitute a part of the total volume of commercial bank liquidity in the Eurobanking system. The amounts of the particular currencies in which this liquidity is denominated are not subject to any special limitation other than those imposed by governments to control the Eurocurrency operations of their commercial banks. In the absence of controls, depositors can acquire Eurodollars or other Eurocurrencies or the domestic currency of the bank receiving the deposit with any convertible currency. The Eurobanks offer interest rates for various Eurocurrencies in accordance with the demand for loans in these currencies, the rate to individual borrowers reflecting the interbank rate for the individual Eurocurrencies. Interest rate differentials between deposits or loans in different Eurocurrencies reflect both speculative forces in the exchange market and domestic credit conditions in the countries whose currencies are involved. Covered Eurocurrency rates tend to move rather closely together, but interest rates for Eurodollars, Euro-Deutsche marks, Euro-Swiss francs, and Eurosterling will frequently depart rather sharply from interest rates in the domestic market of the countries whose currencies are involved.[2] Such deviations generally reflect national monetary controls that seek to insulate the domestic money markets from the Eurocurrency markets.

The constraint on the expansion of Eurocurrency deposits and loans comes not from the supply of dollars or any other currency (which currencies the banks can readily acquire in accordance with their need for meeting obligations) but from the constraints on the total creation of deposit liabilities imposed in the aggregate by the individual actions of the world's monetary authorities. Commercial banks are usually required to keep legal reserves against domestic deposit liabilities, but prudent management would seem to determine that they maintain minimum liquid reserves against all deposit liabilities. However, liquid reserves against external liabilities can be held in a variety of forms and in any convertible currency. What is important for banks is that their liabilities and assets in any particular currency be reasonably well matched by maturity dates. This is a primary function of the interbank market for Eurocurren-

2. For a good discussion of these relationships, see BIS, *Forty-Second Annual Report*, Basle, June 1972, pp. 159–164.

cies; the rates in these markets tend to equalize collective demand and supply for each Eurocurrency and for each maturity. In addition, banks operate in the forward markets for each of the currencies in which they deal.

In analyzing the generation of Eurodollar deposits, it is important to note that not only can nonbank depositors acquire Eurodollar deposits with any convertible currency, but also that the volume of Eurodollar loans does not depend upon the volume of nonbank Eurodollar deposits. Thus, we showed in Tables 2.5 and 2.8 that in 1970 Eurodollar loans to foreign nonbanks approximately doubled while foreign nonbank deposits were nearly unchanged, and that in 1971 Eurodollar loans to foreign nonbanks continued to rise while foreign nonbank Eurodollar deposits declined sharply. The repayment of Eurodollar loans by U.S. residents, the substantial increase in Eurodollar deposits of foreign official institutions, and the generation of Eurodollar loan funds by foreign commercial banks through swap agreements with their central banks or by acquiring U.S. dollars in the exchange market all played a role in these developments. Some of the dollars placed in Eurodollar deposits were used by the Eurocurrency banks for switching into other currencies for loans to customers, or the customers themselves frequently switched the dollars into another currency required by the purpose for which the loan was negotiated. All these developments are scarcely in accord with a model that regards the volume of Eurodollar deposits as determined by the redepositing of Eurodollar loans and the expansion of Eurodollar deposits as requiring the acquisition of American dollar reserves.

Continued U.S. dollar deficits have served to increase the supply of funds for deposits in the Eurocurrency market. This is true because, in the absence of restraining actions by the monetary authorities, countries with balance-of-payments surpluses (whether with the United States or other countries) tend to generate internal as well as external liquidity. The surplus liquidity in the hands of foreign private entities tends to flow into the Eurocurrency market and the growth of reserves of central banks makes them willing to supply dollars to the market. Whether there is an actual loss of reserves for any particular foreign country depends on whether there are net short-term capital transfers out of that country to other countries. If there were no U.S. borrowing from the Eurodollar market and all of the dollars deposited in the Eurodollar banks by residents of a particular country were loaned to other residents of that country for financing domestic expenditures, there would be no loss of official reserves in that country arising out of the operations of the Eurodollar market. Thus it may be said that, while U.S. deficits are not necessary for the expansion of the Eurodollar market, the fact that they have contributed to the expansion of the reserves of foreign countries has increased the volume of liquidity available for deposit in the market, and has made monetary authorities willing to permit capital transfers to other countries via the mechanism of the market.

In summary, foreign Eurodollar balances are generated as a consequence of a desire on the part of foreigners to hold their liquid assets in the form of Eurodollars in response to the relatively attractive interest rates (adjusted for the cost of covering in the forward exchange market) that the foreign banks are offering; the rates that the banks are able to offer depend upon the demand for Eurodollar loans. In short, Fritz Machlup's characterization of Eurodollar creation as a "mystery story" appears to derive more from the fictions of the economic analysts than from the facts of the market.[3]

Substitutability Among American Dollars, Eurodollars, and Other Currencies

Liquid asset holders that tend to acquire international (and not simply domestic) assets have the option of holding a variety of short-term assets denominated in various convertible currencies for which there exist international markets. In addition, the Eurocurrency market provides liquid asset holders with an option as to *where* they will hold their liquid assets regardless of the international currency in which the assets are denominated. Thus, sterling deposits may be held in London or in the form of Eurosterling deposits in other countries, just as dollar deposits may be held in the United States or as Eurodollar deposits in foreign countries. There is also a growing market for Euro-Deutsche marks, Euro-Swiss

3. See Fritz Machlup, "Eurodollar Creation: A Mystery Story," *Banca Nazionale del Lavoro Quarterly Review*, No. 94, September 1970.

francs, Euroguilders, etc. There is a fairly high degree of substitutability among all international liquid assets. However, because of the exchange rate risk and cost of conversion associated with the transfer of one currency to another, the degree of substitutability differs between assets denominated in the same currencies but held in different countries from that between assets denominated in different currencies.

Given the necessary data, the analysis of the foreign demand for the several types of liquid assets would be facilitated by the use of a portfolio-adjustment model, according to which asset holders are assumed to hold a certain composition of assets for a given set of yield differentials, risk evaluations, and other preference variables, and to adjust their portfolios with changes in these variables and in the total value of their portfolios. Changes in holdings of any particular asset, say, U.S. CDs, would be related to changes in the variables determining the composition of the asset holdings and to changes in the net worth of the transactor.[4] Such a model is more realistic than the older short-term capital-flow approach that assumes that, with a given interest differential, investors will continue to shift indefinitely from lower yielding assets to the higher yielding assets.[5] As will be noted below, on the basis of the limited data

4. Examples of the employment of portfolio-adjustment models include H. G. Grubel, "Internationally Diversified Portfolios," *American Economic Review*, December 1968, pp. 1299–1314; and Ralph C. Bryant and P. H. Hendershott, *Financial Capital Flows and the Balance of Payments of the United States: An Exploratory Empirical Study*, Princeton Essays in International Finance, No. 25, Princeton, N.J.: Princeton University Press, June 1970. For an analysis of the problems relating to the formulation of portfolio-adjustment models and a critical review of models of international capital movements, see Edward E. Leamer and Robert M. Stern, "Problems in the Theory and Empirical Estimation of International Capital Movements," *International Mobility and Movement of Capital*, New York: NBER, U-NB 24, 1972, pp. 171–206.

5. Empirical studies of capital movements based on capital-flow models have tended to yield unsatisfactory and conflicting results. Examples of short-term capital-flow models of this type are found in P. W. Bell, "Private Capital Movements and the U.S. Balance of Payments Position," *Factors Affecting the U.S. Balance of Payments*, Washington, D.C.: Joint Economic Committee Compendium of Papers, 1962; and in P. B. Kenen, "Short-term Capital Movements and the U.S. Balance of Payments," and Benjamin J. Cohen, "A Survey of Capital Movements and Findings Regarding their Sensitivity," both in *The United States Balance of Payments*, Washington, D.C.: Hearings before the Joint Economic Committee, 1963.

available we do not find liquid asset holders to be highly sensitive to shifts in relative yields. Each group of liquid asset portfolio holders or transactors has a unique preference function in terms of rate of return, risk and other preference variables, and a relevant net worth variable. For example, U.S. firms operating abroad tend to have a high preference for Eurodollars over American dollars. Moreover, foreign subsidiaries of U.S. firms may be constrained by the U.S. foreign direct investment control program from shifting funds to their U.S. parents that they expect to use later in their foreign operations. However, they may be led to shift these funds into a nondollar currency which they will require for future expenditures if that currency is expected to appreciate in relation to the dollar. (They may even take an uncovered position in such a currency.) Other groups of liquid asset portfolio holders may have quite different preferences, although all portfolio holders are influenced in some degree by changes in relative yields on different types of assets.

Unfortunately, the data available on non-U.S. resident holdings of American dollars, Eurodollars, and other currencies are not disaggregated by class of transactor in a manner that would permit us to formulate and test portfolio-adjustment functions. Nor are there data on changes in the relevant net worth of various categories of transactors. Therefore, we shall not attempt the construction of a comprehensive model of the foreign demand and supply of liquid dollar assets. We shall, however, explore some of the factors determining the composition of foreign liquid asset holdings which may throw light on the foreign demand for various categories of liquid dollar assets. Keeping in mind our underlying concept of a portfolio-adjustment model, we shall consider the evidence relating to the degree of substitutability among American dollars, Eurodollars, and other currencies.

FOREIGN NONBANK HOLDINGS—AMERICAN DOLLARS AND EURODOLLARS

There is a high degree of substitutability between American liquid dollar assets and Eurodollars, just as there is a high degree of substitutability among various types of American liquid dollar assets differentiated by yield, maturity, risk, and cost of conversion into transactions media, i.e., demand deposits in a U.S. resident bank. Although Eurodollar deposits are not generally used for transactions purposes, call Eurodollar deposits with the London branch of a U.S. bank must be as liquid as time deposits in the bank's head office in New York. Eurodollar deposits may involve some additional risk in the minds of holders over direct deposits in U.S. banks or other liquid American assets, but this risk varies a great deal with the country in which the Eurodollar deposit is held. The reason, presumably, is that much of the risk arises from the possibility of government exchange controls restricting the right of the Eurodollar depositor to obtain American dollars.[6] This risk is slight in most Western European countries, even though there are exchange restrictions on the acquisition of Eurodollar deposits by residents in some of these countries.

Foreign holders of liquid dollar assets may find it more convenient to hold Eurodollar deposits with their own bank or with a bank in another European country with which they have close business relations than to hold liquid dollar assets in a bank in the United States which they do not know. They may find it less risky to hold their dollars in a Swiss bank as compared with a U.S. bank if they want to conceal their foreign exchange holdings from domestic authorities. They may even have more confidence in the solvency of a European bank with which they are familiar than in a U.S. bank which they do not know. Thus, it is not always the case that foreigners prefer American dollars over Eurodollars; the preference may be just the opposite. But for U.S. resident liquid asset holders it will nearly always be more convenient and less risky to hold their dollars in the United States, unless, of course, they have tax or other reasons to conceal their assets. However, as noted above, U.S. nonfinancial corporations are subject to a special constraint on holding Eurodollar deposits or any other foreign assets by reason of U.S government controls on the outflow of capital.

6. Historically there have been small differences in rates paid by foreig branches of American banks located in different countries. These interest differ entials may reflect in part differences in risk in the minds of depositors.

Interest Rate Differentials. Prior to 1969, the differential between U.S. CD rates and Eurodollar deposit rates tended to be well under 1 percent and occasionally less than 0.5 percent in favor of Eurodollar deposits. There was a noticeable exception in the second half of 1966 when tight credit conditions in the United States, coupled with a ceiling of 5.5 percent on CDs (issued under Federal Reserve Regulation Q), led U.S. banks to increase their borrowing from the Eurodollar market by some $2 billion between the end of June 1966 and the end of December of that year. This resulted in a differential of 150 basis points (i.e., 1.5 percent) between the three-month Eurodollar rate and the three-month (new issue) CD rate, and a differential of over 100 basis points between the three-month Eurodollar rate and the secondary market yield for ninety-day CDs. During 1969 there was a similar but more striking development as large U.S. bank borrowings from the Eurodollar market forced the three-month Eurodollar deposit rate to 11.3 percent in September 1969, some 530 basis points above the 6.0 percent ceiling on new issue CDs and 256 basis points above the U.S. secondary market yield for ninety-day CDs.[7] By the end of July 1970, the spread between the three-month Eurodollar rate and the U.S. secondary market yield for CDs had narrowed to 40 basis points[8] (see Table 3.1).

During the periods of monetary crisis in 1971 the spread between the Eurodollar deposit rate and the U.S. secondary market CD rate again widened; it was 244 basis points at the end of May 1971 and 313 basis points at the end of August. These large differentials were evidently associated with a heavy demand for Eurodollar loans, partly for conversion into Deutsche marks and other strong European currencies, while the foreign supply of Eurodollar deposits

---

7. For a discussion of the relationship between Eurodollar deposit rates and U.S. money market rates during this period, see Ira O. Scott, Jr., *The Eurodollar Market and Its Public Policy Implications,* Washington, D.C.: Joint Economic Committee, Congress of the United States, 1970.

8. Effective June 24, 1970, maximum interest rates on CDs of $100,000 or more were removed for both U.S. residents and foreign nonofficial residents. Regulation Q has not applied in recent years to CDs in U.S. banks held by foreign official agencies. For this reason, interest rates on official agency CDs (which are subject to negotiation and whose rates are not published) have tended to move within a much narrower range of Eurodollar deposit rates than have CDs available to nonofficial agencies and individuals.

66   *The Behavior of Foreign Dollar Holdings*

TABLE 3.1

Eurodollar Deposit Rates, U.S. Interest Rates, and Changes
in U.S. Domestic and Foreign Branch Bank
Liabilities to Foreign Nonbanks, June 1968–December 1971
(end of month; millions of dollars)

	Three-Month Eurodollar Deposit Rate (1)	U.S. Secondary Market Rate for CDs (2)	(1) − (2)	U.S. Interest-Earning Short-Term Liabilities to Foreign Nonbanks	Change from Previous Month	U.S. Foreign Branch Bank Dollar Liabilities to Foreign Nonbanks	Change from Previous Month
1968							
June	6.75%	6.03%	.72%	2,478		2,101	
July	6.19	5.88	.31	2,498	20	2,248	147
August	6.13	5.85	.28	2,545	47	2,248	0
September	6.19	5.65	.54	2,562	17	2,280	32
October	6.63	6.03	.60	2,601	39	2,455	175
November	6.88	6.08	.80	2,670	69	2,521	66
December	7.06	6.58	.48	2,647	−23	2,538	17
1969							
January	7.56	6.45	1.11	2,676	29	2,447	−91
February	8.38	6.65	1.73	2,620	−56	2,759	312
March	8.44	6.65	1.79	2,638	18	2,924	165
April	8.44	6.85	1.59	2,618	−20	3,220	296
May	10.25	7.55	2.70	2,613	−5	3,282	62
June	10.50	8.25	2.25	2,498	−115	3,798	516
July	10.38	8.75	1.63	2,457	−41	4,528	730
August	11.13	8.25	2.88	2,417	−40	4,734	206
September	11.31	8.75	2.56	2,290	−127	4,748(4,214)	14
October	9.75	8.50	1.25	2,233	−57	4,625	411
November	10.94	8.75	2.19	2,186	−47	4,504	−121
December	10.13	9.00	1.13	2,230(2,352)	44	4,851	347
1970							
January	9.56	8.70	.86	2,312	−40	5,016	165
February	9.31	8.63	.68	2,226	−86	4,932	−84
March	8.50	6.75	1.75	2,198	−28	4,953	21
April	8.56	7.75	.81	2,189	−9	4,900	−53
May	9.06	8.04	1.02	2,279	90	5,278	378
June	9.00	8.13	.87	2,327	48	4,964	−314
July	8.38	7.98	.40	2,279	−48	4,905	−59
August	8.06	7.73	.33	2,258	−21	5,056	151
September	8.38	7.39	.99	2,260	2	4,936	−120
October	7.63	6.65	.98	2,301	41	4,843	−93
November	7.19	5.92	1.27	2,296	−5	5,177	334
December	6.44	5.59	.85	2,348(2,350)	52	4,874	−303

TABLE 3.1 (cont.)

	Three-Month Eurodollar Deposit Rate (1)	U.S. Secondary Market Rate for CDs (2)	(1) – (2)	U.S. Interest-Earning Short-Term Liabilities to Foreign Nonbanks	Change from Previous Month	U.S. Foreign Branch Bank Dollar Liabilities to Foreign Nonbanks	Change from Previous Month
1971							
January	5.81%	4.84%	.97%	2,349	1	4,513	−361
February	5.44	4.21	1.23	2,376	27	4,749	236
March	5.31	3.83	1.48	2,351	−25	4,794	45
April	6.25	4.72	1.53	2,323	−28	4,612	−182
May	7.56	5.12	2.44	2,304	−19	4,630	18
June	6.50	5.43	1.07	2,197	−107	4,775	145
July	6.69	5.80	.89	2,198	1	4,530	−245
August	8.88	5.75	3.13	2,155	−43	4,956	426
September	7.75	5.66	2.09	2,068	−87	4,752	−204
October	5.94	5.18	.76	2,029	−39	4,878	126
November	6.44	4.89	1.55	2,053	24	4,910	32
December	5.75	4.58	1.17	2,031(2,034)	−22	4,953	43

SOURCES: *Treasury Bulletin*, November 1970; *Federal Reserve Bulletin*, various issues; *U.S. Financial Data*, Federal Reserve Bank of St. Louis, various issues; and Morgan Guaranty Trust Company of New York, *World Financial Markets*, various issues.
NOTE: Figures in parentheses are comparable with those shown for the following dates. Figures not in parentheses are comparable with those shown for previous dates.

was restricted by reason of the expectation of a dollar depreciation.[9] But by the end of May 1972, the rate differential in favor of Eurodollar deposits had nearly disappeared.

The effect of these changes in the relative yields on Eurodollar deposits and U.S. CDs on foreign nonbank holdings of American interest-earnings dollar assets and Eurodollar deposits is not readily discernible from the available data. If there were only these two

9. Since there are no legal restrictions on shifting noncorporate funds by U.S. residents to the Eurodollar market (and indeed large amounts of both corporate and noncorporate funds were shifted abroad during 1971), it is somewhat surprising that these rate differentials reached the levels they did. Perhaps large U.S. investors who were able to move their funds around the world moved them into European currencies where their expectations of short-term capital gains were much greater than the gains from shifting funds into Eurodollars.

categories of liquid asset holdings, a priori reasoning would suggest that an increase in the spread between Eurodollar deposit rates and the U.S. CD rate in favor of Eurodollar deposits would be accompanied by a reduction in foreign nonbank holdings of interest-earning American liquid dollar assets in relation to foreign nonbank holdings of Eurodollar deposits. Table 3.1 shows, for the period from June 1968 to December 1971, monthly changes in these holdings, the series on Eurodollar deposits being of necessity limited to the liabilities reported by foreign branches of U.S. banks. Table 3.1 also gives the spread in interest rates at the end of each month over the period. During 1969 the interest rate differential in favor of Eurodollar deposits was substantially greater than in 1968. Over the same period, foreign nonbank deposits in foreign branches of U.S. banks nearly doubled from $2.5 billion to $4.9 billion, while short-term interest-earning American dollar assets held by foreign nonbanks declined from $2.6 billion to $2.2 billion. During 1970 the interest differential, while much lower than in 1969, remained close to the level reached at the end of that year. There was also little overall change in foreign nonbank holdings of either liquid dollar assets in the United States or of Eurodollar deposits in foreign branches of U.S. banks. During the next seven months, to the end of July 1971, there was a decline in foreign nonbank holdings of both American liquid dollar assets and Eurodollar deposits, though the interest differential in favor of the latter was higher on the average. This development undoubtedly reflected the heavy speculation against the dollar during this period. Foreign nonbank holdings of Eurodollar deposits in foreign branches of U.S. banks then recovered abruptly to approximately the level at the end of 1970 as the interest differential again increased sharply, if only temporarily, while foreign nonbank holdings of American liquid dollar assets declined further. During the first four months of 1972, foreign nonbank holdings of Eurodollar deposits with U.S. branches rose sharply to an all-time high of $5.9 billion at the end of April 1972, while their holdings of interest-earning American liquid dollar assets remained approximately the same. Yet the spread between U.S. money market rates and Eurodollar deposit rates in favor of the latter narrowed substantially during this period as compared with 1971.

On a month-to-month basis for the period from June 1968 to December 1971, changes in the composition of liquid dollar assets of foreign nonbanks did not occur in accordance with changes in the spread between the Eurodollar deposit rate and the U.S. secondary market rate for CDs. In 42 observations recorded in Table 3.1, the movements were in accordance with a priori expectations in only 10 cases. This may have been due in part to the nature of the data: only month-to-month changes in assets and end-of-the-month interest rates were available. Weekly data on changes in holdings of liquid dollar assets in relation to the average weekly spread between Eurodollar deposit rates and U.S. CD rates might have indicated greater sensitivity. There are also undoubtedly lags in response to changes in interest rate differentials on the part of portfolio holders. More importantly, however, foreign liquid asset portfolio holders have the option of changing their holdings of non-dollar currency assets in response to interest differentials and to expectations with respect to the exchange value of the dollar. The period under examination was one of heavy speculative movements.

Two other factors might be mentioned in connection with the interest sensitivity of liquid asset holdings. The first is that once the Eurodollar deposits have been acquired, say, with a maturity of three to six months, the depositor must pay a penalty to liquidate them and may, therefore, prefer to hold them to maturity unless there are compelling reasons of risk or speculative advantage to shift to another currency. The second point, stressed in the portfolio-adjustment model of capital flows, is that when investors have shifted their funds in response to a change in yield differentials, a further increase in the differential in favor of a particular type of asset may not bring the same type of response since investors may have already shifted those funds that can be readily shifted without penalty or without interference with some other objective in their preference function.

Other Evidence of Substitution. Although there is little evidence of substitution between foreign holdings of American liquid dollar assets and Eurodollar deposits in response to interest rate differentials, the growth of the Eurodollar market may very well have had a retarding effect on the growth of foreign holdings of American liquid dollar assets. If we consider first foreign commercial banks, satisfactory evidence of such substitution is difficult to uncover because of the large volume of intra-multinational bank balances and the failure of the statistics to distinguish these balances and other foreign commercial bank assets. For foreign nonbank holdings, however, there is a certain amount of indirect evidence of substitution of Eurodollars for American liquid dollar assets. This substitution is probably long run or structural rather than short run in the sense of month-to-month or quarter-to-quarter shifts in holdings between the two types of liquid dollar assets, although we did find a modest inverse correlation between quarter-to-quarter changes in foreign nonbank holdings of American liquid dollar assets and foreign nonbank deposits in foreign branches of U.S. banks over the 1965–70 period.[10]

Between the end of 1957 and the end of 1964, American liquid dollar assets of foreign nonbanks rose by 41 percent, while world trade in current dollars increased by about 50 percent. During this same period, total liquid dollar asset holdings of foreign nonbanks, including both American dollars and Eurodollars, increased by nearly 130 percent (Table 3.2). Between the end of 1964 and the end of 1968, American liquid dollar holdings of foreign nonbanks rose by about 32 percent while world trade in current dollars rose by about 40 percent. During this same period, total liquid dollar assets of foreign nonbanks increased by about 95 percent. However, during the 1968–70 period, American liquid dollar asset holdings of foreign nonbanks declined, but world trade increased by about 30 percent in current dollars. It was during this period that foreign nonbank holdings of Eurodollars achieved their maximum growth, rising from $9.3 billion at the end of 1968 to $17.6 billion at the end of 1970. Thus, in spite of a decrease in foreign nonbank holdings of American dollars, total liquid dollar holdings of foreign nonbanks increased by 56 percent over the 1968–70 period. In 1971, foreign nonbank holdings of both U.S. dollars

10. When we regressed quarter-to-quarter changes in foreign nonbank holdings of American liquid dollar assets on quarter-to-quarter changes in foreign nonbank dollar deposits with foreign branches of U.S. banks, we obtained a significant inverse correlation but the R^2 was only .19. (Both the slope and coefficient of correlation are significantly different from zero at the 99 percent level of confidence.)

TABLE 3.2

World Trade and Liquid Dollar Holdings
of Foreign Nonbanks
(end of period; billions of dollars)

	1957	1964	1968	1969	1970	1971
World trade (imports, cif)[a]	106.8	161.2	225.0	256.4	294.1	329.0
Foreign nonbanks						
1. U.S. dollars[b]	2.7	3.8	5.0	4.6	4.7	4.2
2. Eurodollars[c]	0.5	3.5	9.3	17.2	17.6	14.5
Subtotal	3.2	7.3	14.3	21.8	22.3	18.7

a. *International Financial Statistics*, various issues.
b. Table 2.1.
c. Table 2.5. The figure for 1964 is based on BIS data. The figure for 1957 is derived from dollar liabilities to nonbanks reported by the Bank of Canada and the Bank of England.

and of Eurodollar deposits declined, probably as a consequence of the uncertainty regarding the future exchange value of the dollar. While by no means conclusive, the above data strongly suggest substitution of Eurodollar deposits for American liquid dollar holdings from the end of 1968 to the end of 1970.

With regard to foreign official holdings of Eurodollar deposits, their rapid growth (especially during 1970 when they rose from $4.3 billion to $9.2 billion) suggests that foreign official institutions were substituting Eurodollar deposits for American liquid dollar holdings (Table 2.6). Undoubtedly there was a substantial net substitution on the part of central banks and governments outside of the major Western European countries and Japan. However, as has already been explained, the depositing of American dollars by foreign central banks in the Eurodollar market does not reduce the American dollar holdings of foreign official institutions as a group. Such depositing increases the total liquid dollar holdings of foreign official institutions.

FOREIGN NONBANK HOLDINGS—EURODOLLARS
AND NONDOLLAR CURRENCIES

In recent years foreign nonbank holders of liquid asset portfolios have held the bulk of their liquid dollar assets in the form of Euro-

72   *The Behavior of Foreign Dollar Holdings*

dollar deposits.[11] In fact, foreign nonbank holdings of Eurodollar deposits with "inside area" and Canadian banks alone were nearly six times the level of their holdings of short-term, interest-earning U.S. dollar assets at the end of 1969 and at the end of 1970. Foreign nonbank holdings of American short-term, interest-earning dollar assets (the sum of lines IC2 and IC3 in Table 2.1) have been remarkably stable, at least since 1965, ranging between $2 and $3 billion, while there have been substantial fluctuations in their holdings of Eurodollars. Thus it is unlikely that the rise in foreign nonbank holdings of Eurodollars from $7.0 billion at the end of 1966 to $17.6 billion at the end of 1970 (Table 2.5) occurred mainly at the expense of their holdings of American liquid dollar assets, although it is likely that there was some substitution. Nor can the sharp drop of $3.1 billion in foreign nonbank Eurodollar holdings between the end of 1970 and the end of 1971 be explained by a shift into American liquid dollar assets (holdings of which also declined). Thus fluctuations in the volume of Eurodollar deposits held by foreign nonbank portfolio holders must be largely explained by substitution between Eurodollars and other currencies rather than between Eurodollars and American dollars.

As we have already observed, prior to 1969 the spread between interest rates on Eurodollar deposits and comparable U.S. money market rates was usually less than 100 basis points. This was also true during 1970. Only during abnormal money market or foreign exchange market conditions has this spread risen above 200 basis points. On the other hand, uncovered interest rate differentials between Eurodollar deposit rates and money market rates in certain countries, e.g., Germany and Switzerland, were continuously in excess of 200 basis points prior to 1969 and differentials of 300–400 basis points have not been uncommon.[12] Moreover, as may be noted in Table 3.3, substantial differentials between *covered* Eurodollar deposit rates and comparable money market rates in Britain, Germany, and Switzerland, ranging well above 300 basis points, have

11. Except for working balances of trading firms, Eurodollar deposits probably constitute half or more of all foreign nonbank liquid asset holdings in currencies other than the domestic currency of the holder.

12. Differentials of over 500 basis points between uncovered three-month Eurodollar deposit rates and Swiss three-month deposit rates occurred in 1969, 1971, and 1972 and between Eurodollar deposit rates and German three-month deposit rates in 1969.

TABLE 3.3

Quarterly Changes in Foreign Nonbank Dollar Deposits in Foreign Branches of U.S. Banks, and Covered Eurodollar–Domestic Market Interest Rate Differentials (percentages)

Quarters	Quarterly Change in Nonbank Dollar Deposits in Foreign Branches of U.S. Banks		Quarterly Average Covered Eurodollar-Domestic Market Interest Rate Differential		
	All Branches	U.K. Branches	U.K.[a]	Swiss[b]	German[c]
	(1)	(2)	(3)	(4)	(5)
1966					
3rd			.45	1.46	.87
4th	11.24	13.11	.33	2.01	.53
1967					
1st	−4.07	−5.80	−.04	.73	.87
2nd	4.31	.67	.12	.64	.53
3rd	12.13	17.62	.31	.35	.45
4th	13.16	18.68	.95	.95	−.33
1968					
1st	4.96	1.63	1.79	.49	.13
2nd	3.45	4.60	3.38	.94	.29
3rd	8.52	14.67	1.05	.65	.33
4th	11.32	13.03	2.11	1.62	−.47
1969					
1st	15.21	14.44	2.66	1.88	.13
2nd	29.89	31.64	5.40	2.96	−.40
3rd	25.03	23.43	6.37	4.66	.17
4th	15.12	7.48	2.14	2.90	.95
1970					
1st	2.10	2.40	.64	3.44	.37
2nd	.22	−3.22	1.21	1.91	.31
3rd	−.56	−5.60	1.57	.74	−.01
4th	−1.26	−.93	.85	.61	−.32
1971					
1st	−1.64	−5.03	.83	.41	−.27
2nd	−.40	−5.20	1.59	.93	−.23
3rd	−1.15	−.78	1.59	−.05	−1.14
4th	4.23	4.60	.89	−1.09	−1.37

SOURCES: Morgan Guaranty Trust Company of New York, *World Financial Markets*, and IMF, *International Financial Statistics* for interest rates. Foreign branch data from *Federal Reserve Bulletin*, February 1972 and July 1972.

a. Covered Eurodollar three-month deposit rate minus three-month U.K. local authority deposit rate.

b. Covered three-month Eurodollar rate minus three-month Swiss deposit rate.

c. Covered Eurodollar deposit rate minus three-month German deposit rate.

74   *The Behavior of Foreign Dollar Holdings*

Regression Analysis of Data from Table 3.3

	R^2	F-test on Cor. Coef.	t-test on Variable	Degrees of Freedom	Durbin-Watson
1. Col. (1) on col. (3)	.50	19.1ª	sig. at 1% level	19	1.15
2. Col. (2) on col. (3)	.34	9.6ª	sig. at 1% level	19	1.16
3. Col. (1) on col. (4)	.38	11.7ª	sig. at 1% level	19	.94
4. Col. (2) on col. (4)	.23	5.5ª	sig. at 5% level	19	.91
5. Col. (1) on cols. (3) & (4)	.55	11.2ª	(3) sig. at 5% level (4) sig. at 1% level	18	1.2
6. Col. (2) on cols. (3) & (4)	.36	5.0ᵇ	(3) sig. at 5% level (4) not sig.	18	1.14
7. Col. (1) on col. (5)	no rel.			19	
8. Col. (2) on col. (5)	no rel.			19	

Cols. (1) + (2) D-W indeterminate at both 5% and 1% levels
Cols. (3) + (4) D-W significant at 1% level
Cols. (5) + (5) D-W significant at 5% level

NOTE: Due to the possible presence of autocorrelation, both the values and the significance of the R^2s may be greatly overstated.
  a. Significant at the 1% level.
  b. Significant at the 5% level.

occurred at times in recent years. These differentials reflect the risk of exchange controls, transactions costs, and governmental restrictions on capital movements. Nevertheless, the covered rate differentials in particular appear at times to be surprisingly large. The extent to which foreign nonbank depositors cover their Eurodollar deposits is not known to the authors. Conceivably many liquid asset holders are motivated more by a desire to speculate on a change in exchange rates than by a desire to earn a larger interest yield during certain periods.

The only reasonably reliable monthly or quarterly data on foreign nonbank Eurodollar deposits are those for deposits in foreign branches of U.S. banks. Again there is no breakdown of these deposits by country of origin. Nevertheless, we sought to determine whether there was any relationship between changes in foreign nonbank deposits and changes in differentials between covered three-month Eurodollar deposit rates and domestic three-month deposit rates in Britain, Switzerland, and Germany, respectively. We regressed month-to-month changes in foreign nonbank deposits in foreign branches of U.S. banks and in U.K. branches alone on the average monthly differentials between covered Eurodollar deposit rates and domestic deposit rates in Britain, Switzerland, and Germany, respectively, over the period June 1966–December 1971. We found significant coefficients for our regressions but rather low R^2s (on the order of .15). We tried lagging the monthly data but with no better results. We then did the same regressions employing quarterly data for the period June 1966–December 1971 and found some significant relationships (see Table 3.3). For the regression of percentage (quarter-to-quarter) changes in foreign nonbank deposits in all foreign branches of U.S. banks on the covered Eurodollar-U.K. Local Authority deposit rate differential, we obtained an R^2 of .50. For the same regression using the covered Eurodollar-Swiss deposit rate differential, we obtained an R^2 of .38. (In both cases the F-test on the correlation coefficient and the t-test on the variable were significant at the 1 percent level but the DW statistic suggests the possible presence of autocorrelation.) When we regressed percentage (quarter-to-quarter) changes in foreign nonbank deposits in all foreign branches of U.S. banks on the Eurodollar-U.K. Local Authority deposit rate and the Eurodollar-Swiss deposit rate differentials combined, we obtained an R^2 of .55. However, we found no relation using the covered Eurodollar-German deposit rate differentials. We also performed the same regressions using foreign nonbank deposits in U.K. branches of U.S. banks. The results were significant (again except for the covered Eurodollar-German deposit rate differential), but the R^2s were lower. While this evidence is by no means conclusive, it does point to the existence of a strong influence of interest rate differentials on the behavior of foreign nonbank Eurodollar deposits.

CHANGES IN EURODOLLAR POSITIONS
OF FOREIGN COMMERCIAL BANKS

Eurodollar banks usually stand ready to accept all Eurodollar deposits offered, but the rate of interest they are willing to pay depositors depends upon the demand for Eurodollar loans from Eurodollar banks as a group. An individual Eurodollar bank can redeposit with other Eurodollar banks any excess of deposits over the demand for Eurodollar loans. When the demand for loans exceeds Eurodollar deposits, a Eurodollar bank can obtain additional funds in the inter-Eurodollar bank market, or from the foreign exchange market, or from its central bank under a swap arrangement or outright purchase against domestic currency. The demand for Eurodollar loans will depend in considerable measure on the relationship between Eurodollar loan rates and the loan rates in the domestic money markets. Large well-known firms can usually obtain Eurodollar loans from the cheapest source by contacting Eurodollar banks in different countries.

Since the Eurodollar market for both loans and deposits is highly competitive, deposit rates and loan rates tend to be maintained within a rather narrow range within the Eurodollar banking system. Eurodollar loans are in competition with loans from domestic market sources in individual countries, although in some countries local borrowers are restricted by the monetary authorities as to how much and under what conditions they can borrow in the Eurodollar market. Borrowers requiring local currency for working capital must bear the exchange risk of repaying dollars or other borrowed Eurocurrencies unless they cover their position in the forward market. Alternatively, the Eurocurrency banks may convert dollars or other currencies obtained from Eurocurrency deposits into domestic or third currencies for making loans. In this case, the banks must either cover their positions in the same currencies or bear the exchange risk if their Eurocurrency liabilities exceed their Eurocurrency assets in a particular currency.

Ultimately the interest paid on Eurodollar deposits is determined by the demand for Eurodollar loans. During 1968 and 1969, when U.S. banks were borrowing heavily in the Eurodollar market, the U.S. demand determined in large measure the interest rates on Eurodollar deposits. However, since the repayment of U.S. resident indebtedness in 1970 and 1971, the rates of interest that Eurodollar banks have been willing to offer for Eurodollar deposits have been governed largely by the foreign demand for loans, which demand in turn is affected by domestic money market rates together with government regulations on borrowing from the Eurodollar market.

In a recent study, Rodney H. Mills[13] examined the relationship between net Eurodollar positions of commercial banks in Belgium, the Netherlands, France, and Germany on the one hand, and covered differentials between three-month Eurodollar rates and selected domestic money rates on the other. Using quarterly BIS data for the period September 1963–June 1969, Mills formulated a model for explaining changes in the net Eurodollar positions of the commercial banks in each of these four countries with nonresidents of the countries in which the banks were located. His independent variables were (1) the average covered differential between the three-month Eurodollar rate and a selected domestic money market rate in the quarter or in the month preceding the end of the quarter; (2) bank loans to private domestic borrowers; and (3) for Belgium and Germany, certain other variables. His multiple regression analysis shows that changes in covered interest differentials were closely associated with changes in commercial bank net Eurodollar positions vis-à-vis nonresidents, excluding U.S. residents, for each of the four countries. One explanation for his results is that commercial banks find it profitable to borrow in the Eurodollar market for expanding loans to their customers when the covered Eurodollar interest rate is lower than the domestic market rate, and to lend to the Eurodollar market (through depositing dollars in other Eurodollar banks) when the covered Eurodollar rate exceeds the domestic market rate. This explanation reflects a close relationship between domestic lending and the Eurodollar market when there is no substantial interference by government regulations. Since such regulations existed during the period under examination in the United Kingdom and Italy, these countries had to be excluded from his analysis. Also, since the large BIS operations are included in the BIS data on the dollar position of Swiss banks vis-à-vis nonresidents, Mills found it necessary to exclude Switzerland from his analysis.

Mills' study constitutes an important contribution to the explanation of changes in net dollar positions vis-à-vis nonresidents

13. See Rodney H. Mills, Jr., *Explaining Changes in Eurodollar Positions: A Study of Banks in Four European Countries,* Discussion Paper No. 1, Washington, D.C.: Federal Reserve Board, Division of International Finance, August 27, 1971.

(excluding U.S. residents) for the commercial banks of the four countries covered. However, the task of explaining the aggregate demand and supply of funds for the Eurodollar market as a whole and of devising a satisfactory model for the determination of interest rates in the market remains a complex and formidable undertaking for which the statistical data available so far are seriously inadequate.

Influence of the Eurodollar Market on the U.S. Balance of Payments

While we have stressed that the Eurodollar market can expand or contract independently of the U.S. balance of payments, there is in practice an interaction between developments in the Eurodollar market and changes in the U.S. balance of payments. As has already been noted, U.S. balance-of-payments deficits may increase the volume of liquidity in the rest of the world, which in turn may expand the volume of Eurodollar deposits. The Eurodollar market has attracted large U.S. private short-term capital flows to the market, which tend to add to the dollar holdings of foreign central banks and thus to have an adverse effect on the U.S. balance on official reserve transactions account. Conversely, when U.S. residents borrow from the Eurodollar market, the U.S. balance on official reserve transactions account will tend to improve. More broadly viewed, however, both the flow of U.S. resident funds to the market and U.S. borrowings from the market may serve to expand the Eurodollar market. Thus the large borrowing by U.S. banks was a major factor in the rapid expansion of the Eurodollar market during 1968 and 1969; the higher Eurodollar deposit rates relative to domestic money market rates abroad attracted foreign depositors, as did also the promotional activities of foreign branches of U.S. banks seeking Eurodollar funds for their parent banks. When the U.S. borrowings were repaid during 1970 and 1971, Eurobanks, including foreign branches of U.S. banks, sought new foreign loan markets in which to place funds flowing from the United States. The "net size of the market" as measured by the BIS continued to expand, although at a somewhat slower pace than in 1968 and 1969.

Over the 1965–70 period, U.S. resident borrowings from and

repayments to the Eurodollar market through foreign branches of U.S. banks were responsible for a substantial portion of the quarter-to-quarter changes in net foreign nonofficial claims on U.S. residents.[14] Since to a substantial degree the dollars flowing out of the Eurodollar market to U.S. residents either came from foreign central banks or served to reduce additions to their American dollar holdings, and since the American dollars repaid to the Eurodollar market tended to flow into foreign central banks, there was also a strong association between changes in net U.S. resident borrowing from the Eurodollar market and changes in the U.S. balance on official reserve transactions account over the 1965–70 period. Thus during the first three quarters of 1968 and the first and second quarters of 1969, when U.S. resident borrowing from the Eurodollar market was quite heavy, the U.S. balance on official reserve transactions account was in surplus in each of these quarters. On the other hand, in 1970 when U.S. residents were making large repayments to the Eurodollar market, the U.S. balance on official reserve transactions account was in substantial deficit in every quarter.[15] In this way, borrowings from the Eurodollar market by U.S. banks served for a time to hide the deterioration in the U.S. balance of payments as far as the effects on foreign official holdings are concerned and then, when the borrowings were repaid, to increase the redundancy of dollars in foreign reserves.

While these U.S. short-term capital movements are directly related to the operations of the Eurodollar market, it has been suggested that, in the absence of the Eurodollar market, foreign short-term capital might have flowed to the United States during 1968 and 1969 in the presence of a tight U.S. money market, with much the same consequences for the U.S. balance of payments. We doubt

14. The results of regression analysis of the relationship between quarterly changes in net foreign nonofficial liquid claims on U.S. residents and in changes in the U.S. official reserve transactions balance, on the one hand, and in the net claims of foreign branches of U.S. banks on U.S. residents, on the other, are presented in an article by Raymond F. Mikesell entitled "The Eurodollar Market and the Foreign Demand for Liquid Dollar Assets," *Journal of Money, Credit and Banking,* August 1972, pp. 643–683.

15. In the second quarter of 1970 there was a small increase in net claims of foreign branches of U.S. banks on U.S. residents ($218 million), but this was not sufficient to offset other factors in the U.S. balance of payments making for a deficit on the U.S. official reserve transactions account.

very much that these flows would have occurred in anything like the same volume. In the first place, Regulation Q, which limited the rates of interest U.S. commercial banks could pay on time deposits and CDs, applied to foreign nonofficial entities as well as to U.S. residents. Hence, to a considerable degree, U.S. commercial bank borrowing from abroad through their foreign branches was a means of circumventing Regulation Q as applied to foreigners. Second, prior to October 1969, U.S. banks were not required to maintain reserves on liabilities to their foreign branches. More broadly, however, we believe that the Eurodollar market provided a mechanism for international financial intermediation for the mobilization of large amounts of foreign liquid funds that would never have been placed directly in the United States even in the absence of Regulation Q.

EFFECTS ON THE BASIC BALANCE

Perhaps a more significant question is whether the growth of the Eurodollar market has had any impact on the U.S. basic balance, either the goods and services component or the long-term capital component. It is impossible to provide satisfactory evidence either way; we can only speculate as to what the causal relationships might be, if any. At the most general level, there have been allegations in earlier postwar years that the growth of world trade was being impaired by a shortage of international liquidity. The argument was usually made with respect to the rate of growth of official reserve assets rather than private international liquidity. The Eurodollar market made a modest contribution to foreign official reserve assets as a consequence of foreign official depositing in the Eurodollar market. This did not become very significant until after 1967, however, and few would argue that during this period world trade had been constrained by a lack of dollar liquidity generally, or by a lack of official reserves.

The expansion of international credit facilities provided by the Eurodollar market may have induced some increase in world trade, but such an expansion would not necessarily contribute to an improvement in the U.S. trade balance. It is possible for U.S. exports to benefit from a redistribution of international liquidity. In the absence of the Eurodollar market, an increase in foreign dollar liquidity generated by a U.S. payments deficit would flow into the reserves of the surplus countries. With the market, some of these dollar funds may flow in the form of credits to deficit countries where they become available for increased imports from the United States and elsewhere. This flow would reduce the volume of domestic credit in the capital exporting countries. In the absence of these flows, the capital exporting countries might have contracted credit as an anti-inflationary measure.

The U.S. current account and the long-term capital account may be affected by the ability of U.S. firms or their foreign affiliates to borrow Eurodollars for financing working capital requirements or for financing capital projects in foreign countries. The latter type of investment frequently takes place under a Eurodollar revolving commitment whereby a Eurodollar bank or a consortium of banks makes a commitment for a period of from, say, three to five years to lend the borrower up to a specified amount. Although actual borrowings are usually evidenced by notes of maturities of less than twelve months, the notes can be renewed during the overall period of commitment, provided agreement is reached on the interest rate for each renewal.[16] Given the U.S. capital export control program, it is quite likely that some U.S. direct foreign investment would not have been made in the absence of the Eurodollar market. On the other hand, some of these investments might have been made by means of direct capital flows from the United States, which flows would have affected the U.S. balance on basic transactions account. To the extent Eurodollar credits have permitted a higher level of U.S. direct investments, such investments may have tended both to increase U.S. exports to U.S. affiliates abroad and to provide substitutes for U.S. exports to other customers.[17]

16. See Morgan Guaranty Trust Company, *The Financing of Business with Eurodollars,* New York, 1967, p. 8.

17. The increase might be expected to come in the form of capital goods during the investment phase and thereafter in the form of materials and components. The substitution effect would come when the affiliate's production displaces goods previously supplied by the parent company. The relationship between U.S. exports and imports and U.S. foreign investment is, however, a controversial issue with which we shall not be further concerned in this book. See G. C. Hufbauer and F. M. Adler, *Overseas Manufacturing Investments and the Balance of Payments,* Washington, D.C.: U.S. Treasury Department, 1968; and Jack N. Behrman, *Direct Manufacturing Investment, Exports, and the Balance of Payments,* New York: National Foreign Trade Council, 1968.

A final consideration with respect to the impact on the U.S. balance of payments has to do with the effects of the Eurodollar market on the international financial intermediation function of the United States. The carrying out of this function has meant that the United States has tended to make extensive loans and direct investments abroad, financed to a substantial degree by the foreign acquisition of liquid dollar balances. The extension of the U.S. banking system abroad has, to a degree, changed this financial intermediation process. Instead of loans being made by the transfer of American dollars from the United States, the loans are made by U.S. banks operating abroad out of funds deposited by foreigners in U.S. branches or in other Eurodollar banks that redeposit the funds with U.S. branches. Insofar as the financing is shifted in this way to foreign-owned dollars, in place of newly provided funds from the United States, the deficit in the U.S. international accounts is reduced. This transfer of the U.S. intermediation function is further facilitated by the development of the Eurobond market, also promoted by U.S. financial institutions, whereby long-term funds required by U.S. firms operating outside the United States are raised abroad. To what extent there has been a transfer of foreign loan financing from parent banks in the United States to their foreign branches is not known to the authors, but the amount must have been considerable. The course of this process has been greatly facilitated by the existence of the U.S. Voluntary Foreign Credit Restriction Program (VFCR), so that it is difficult to determine what the effect of the increased foreign financing by foreign branches of U.S. banks would have been in the absence of those restrictions.

National Monetary Policies and Controls in Relation to the Eurodollar Market

As just indicated, the rapid development of the Eurodollar market after 1963 was in considerable measure a consequence of U.S. credit and capital export controls, including the limits imposed by Regulation Q on the rate of interest U.S. commercial banks could pay on time deposits and CDs, the VFCR program, and the controls on U.S. direct foreign investment. These controls led to a partial insulation of the U.S. capital market from that of the rest of the world, which in turn promoted a market for time deposits denominated in dollars outside the United States and a very large expansion of U.S. foreign branches of U.S. banks to participate in this market. However, while the U.S. domestic money market was to a degree insulated from the Eurodollar market, the Eurodollar market was strongly influenced by U.S. monetary policy and became a channel for the transmission of U.S. domestic monetary conditions to money markets abroad. One effect of the Eurodollar market was to narrow interest rate differentials and to equalize credit conditions among foreign money markets. If a particular country desired to tighten credit, say, by increasing reserve requirements on domestic deposits, the banking system could acquire funds for loans to domestic borrowers by attracting Eurocurrency deposits from abroad. When, in 1971, some European countries began requiring their commercial banks to maintain reserves, or marginal reserves, on liabilities to foreigners, many large domestic corporations simply bypassed the domestic banking system and borrowed abroad. Likewise, when the monetary authorities of a particular country wanted to ease credit conditions, say, by reducing reserve requirements, both commercial banks and nonbanks sought higher yields abroad, particularly through the Eurocurrency market. In addition to interfering with domestic credit objectives, the capital outflow tended to reduce central bank reserves.

Throughout the 1960s some European countries limited direct access to the Eurodollar market by their own residents. This has been true of Britain, France, and Italy, among others. Frequently, as in the case of Britain, residents were limited by exchange control regulations with respect to the acquisition of foreign currencies, and commercial banks were limited in the amount of net foreign assets they could acquire. In recent years, however, regulations have been imposed by countries such as Germany and Switzerland, not for the purpose of preventing short-term capital outflow, but rather for limiting the inflow. The imposition of these controls has arisen in part from the desire to limit the influx of dollars into the central banks and in part from the desire to avoid the credit-expanding effects of the acquisition of additional reserves by domestic banks.[18] For example, the German government imposed marginal reserve requirements on nonresident liabilities of commercial banks in 1971, and in February 1972 it announced a 40 percent deposit requirement on foreign borrowings of nonbanking corporations. Switzerland introduced a 100 percent reserve requirement on net foreign liabilities of banks, banned the sale of domestic securities and certain other financial assets to foreigners, and imposed a 2 percent quarterly tax on foreign deposits held in Swiss banks in excess of amounts held on June 30, 1972.[19]

The effect of these capital import restriction measures by European countries has been to limit the demand for Eurodollar funds arising from firms operating in these countries seeking to obtain credit and from speculators throughout the world that have borrowed Eurodollars for the purpose of converting them into strong European currencies. However, the demand for Eurodollar loans has expanded in other parts of the world, particularly in the developing areas and Eastern Europe.[20] So long as there is sufficient demand for Eurodollars to provide banks and nonbank depositors with rates of return higher than those available in their domestic money markets, funds will flow into the market from Western Europe, the United States, and other areas where banks and other liquid asset holders are permitted to place funds in the market.

18. The question of the ability of domestic monetary authorities to offset the effects of capital inflow on the monetary system has been the subject of a number of studies. See, for example, Manfred Willms, "Controlling Money in an Open Economy: The German Case," *Federal Reserve Bank of St. Louis Review,* April 1971; and Michael G. Porter, "Capital Flows as an Offset to Monetary Policy: The German Experience," *IMF Staff Papers,* July 1972, pp. 395–424. Some analysts have sought to show that capital inflows have not contributed to a net expansion of domestic credit, largely on the grounds that domestic reserves have increased by more than could be accounted for by capital inflows. Others have agreed with monetary authorities of Germany and certain other countries who have held that, in the absence of direct controls over capital inflows, there are severe limits to preventing such flows from increasing the monetary base.

19. For a review of recent monetary controls in European countries and Japan, see Charles A. Coombs, "Treasury and Federal Reserve Foreign Exchange Operations," *Monthly Review,* Federal Reserve Bank of New York, September 1972, pp. 210–232.

20. *Ibid.,* pp. 230–232.

4

Summary and Conclusions

Empirical Findings

HOLDINGS OF AMERICAN LIQUID DOLLAR ASSETS

Foreign nonofficial nonbank holdings of American liquid dollar assets have been relatively stable in recent years; they have varied by less than $1 billion between the end of 1965 and mid-1972 and have exhibited no upward trend over the period. U.S. demand deposit holdings of this group have shown little change, at least since the end of 1963. Given the fact that international trade valued in current dollars has more than doubled since then, the stability of these dollar holdings suggests that foreign nonbanks have found ways of economizing on their dollar balance requirements for both transactions and precautionary purposes. It also indicates that, in view of the several-fold increase in foreign nonbank holdings of other international liquid assets, American dollar balances have become relatively less remunerative or perhaps less desired for other reasons. It should be pointed out that these trends were evident well before 1970 while the dollar was strong in the foreign exchange market.

The behavior of American liquid dollar balances of *foreign commercial banks* is obscured by the data. We have only a rough breakdown between those balances that constitute intra-multinational bank accounting entries and other balances, and the latter are not fully disaggregated by category of liquid dollar asset, e.g., demand deposits. From the end of 1966 to the end of 1970, 68 to 76 percent of all reported U.S. liquid liabilities to foreign commercial banks constituted intra-multinational bank balances. During each of these years except 1970, the largest amount was ac-

85

counted for by the liabilities of U.S. banks to their foreign branches, which rose from $4.0 billion at the end of 1966 to a high of $14.5 billion at the end of November 1969 and then dropped to $6.2 billion by the end of 1970 and to only $1.3 billion by the end of 1971. These balances arose out of U.S. commercial bank borrowings from the Eurodollar market through their foreign branches and were, for the most part, not related to the financing of international transactions. The dollar liabilities of the U.S. branches and agencies of foreign banks to their head offices and affiliates abroad showed a fairly steady growth from $3.3 billion at the end of 1964 to $6.0 billion at the end of 1970. Thereafter they dropped to $4.5 billion by the end of 1971. These liabilities constitute in large measure a commitment of foreign commercial banks to banking operations in the United States concerned with both U.S. domestic and international finance. Given the circumstances under which the large volume of intra-multinational bank balances was created, it is impossible to determine either the amounts of foreign liquid dollar balances which serve an international transactions function or the amounts which represent the American dollar asset components of the liquid interest-earning portfolios of foreign commercial banks.

As long as the rise in foreign liquid claims on the United States was largely in the form of intra-multinational bank balances, the holdings of *foreign official institutions* were fairly stable, the net increase being only $0.2 billion from the end of 1964 to the end of 1969. Then, with the decline in intra-multinational bank balances, foreign official claims rose rapidly. The bulk of the increase in these claims in 1970 ($7.7 billion) could be accounted for by the fall in claims of foreign branches of U.S. banks on their parents ($6.4 billion). In 1971, however, foreign official holdings of American dollars soared by $26.9 billion, of which no more than $5 billion could be attributed to the further liquidation of U.S. commercial bank liabilities to the Eurodollar market. That increase brought total liquid claims on the United States by foreign official institutions at the end of 1971 to just over $50 billion, or more than twice the amount a year earlier and more than three times that of two years earlier.

In contrast to foreign liquid claims on the United States, the recorded) volume of *U.S. liquid claims on foreigners* has not increased substantially since 1964. It was, in fact, slightly higher at the end of 1968 ($10.9 billion) than at the end of 1971 ($10.8 billion). The U.S. *net* liquid position vis-à-vis each of the three categories of foreign liquid dollar asset holders behaved quite diferently over the 1964–71 period. The U.S. net (negative) liquid position with foreign *nonbanks* remained within a range of from $3.2 to $4.8 billion over the period 1963–71 and was $3.8 billion at the end of 1971. However, the U.S. *short-term* position with foreign nonbanks was *positive* throughout the period, taking into account not only liquid items but also U.S. short-term loans and credits to foreigners. (In addition, U.S. corporations had large claims on their foreign affiliates which are generally not included in the data on short-term indebtedness.) The U.S. *net* (negative) liquid position with foreign commercial banks (including foreign branches of U.S. banks) rose from $1.1 billion at the end of 1964 to $16.4 billion by the end of 1969, but declined to $1.4 billion by the end of 1971. The recorded positions, however, grossly understate U.S. resident claims on foreign commercial banks. The United States probably had a net positive position vis-à-vis foreign commercial banks at the end of 1971 if the several billion dollars in (largely unreported) U.S. resident holdings of Eurodollar deposits are included. Thus at the end of 1971, of the three categories distinguished, the United States had a net *negative* short-term position only with foreign official institutions of approximately $50 billion.

HOLDINGS OF EURODOLLARS

Since Eurodollar deposits perform many of the same functions as American dollars, we have estimated the volume of foreign holdings of Eurodollars again by nonofficial nonbanks, commercial banks, and official institutions. Because of the data limitations, the Eurodollar balances of *foreign nonofficial nonbanks* have been limited to their Eurodollar holdings with "inside area" and Canadian banks. On this basis foreign nonbank deposits rose from $7.0 billion at the end of 1966 to $17.6 billion by the end of 1970, but fell to $14.5 billion by the end of 1971. Foreign nonbank borrowings of Eurodollars rose throughout the same period from $5.8 billion at the end of 1966 to $24.8 billion by the end of 1971. The largest increase occurred during 1970, the period of largest repayments of U.S. commercial banks to the Eurodollar market.

Gross Eurodollar holdings of *foreign commercial banks* include some $40 to $50 billion of interbank balances of "inside area" banks and of other foreign commercial banks and foreign branches of U.S. banks active in the market. (Banks are regarded as active in the Eurodollar market if they engage in mutual depositing with one another as contrasted with banks that may simply borrow funds from the market or deposit funds in the market in much the same way as nonbanks.) The BIS excludes all interbank dollar balances of "inside area" banks in its calculation of the net size of the Eurodollar market on the same grounds that interbank balances are excluded in the calculation of domestic money supply in national economies. In our conceptual framework of the Eurodollar market, we regard all foreign commercial banks that are active in the market in the sense indicated above, plus foreign branches of U.S. banks, as constituting the Eurodollar interbank system. However, we regard the foreign branches of U.S. banks as playing a dual role. They are part of both the U.S. banking system and the Eurodollar interbank system. In estimating the Eurodollar deposits of foreign commercial banks in the Eurodollar banking system, we exclude *all* deposits of these banks with one another. However, we include in our estimates of Eurodollar deposits of foreign commercial banks their *net* deposits with foreign branches of U.S. banks. There are, in addition, a large number of foreign commercial banks outside the Eurodollar banking system that have deposits with foreign commercial banks which form a part of the system. Unfortunately, since there is no way by which the volume of these deposits can be derived from the available data, they are excluded from our estimate of Eurodollar deposits of foreign commercial banks. The net position of all foreign commercial banks with foreign branches of U.S. banks rose from $2.2 billion at the end of 1966 to $9.7 billion by the end of 1969. The net position declined to $4.0 billion by the end of 1971 with the repayment of U.S. commercial borrowings to the Eurodollar market.

Foreign official holdings of Eurodollar deposits have become increasingly important in recent years with identified deposits rising from an estimated $1.3 billion at the end of 1964 to $10.1 billion by the end of 1971. Moreover, a portion of the $8.7 billion in unidentified official holdings of Eurocurrencies and other reserve assets are believed to represent Eurodollar deposits.

COMBINED HOLDINGS AND NET POSITIONS

On the basis of the definitions of liquid dollar holdings given above, we have calculated the combined American liquid dollar and Eurodollar holdings of the three categories of foreign dollar holders. Total combined liquid dollar holdings rose from $36.4 billion at the end of 1966 (of which $25.2 billion were American liquid dollars), to $93.5 billion by the end of 1971 (of which $64.9 billion were American liquid dollars). Almost half of the combined foreign liquid dollar holdings at the end of 1966 and almost two-thirds at the end of 1971 represented the claims of foreign official institutions on U.S. residents. The combined liquid dollar holdings of *foreign commercial banks* rose from $8.2 billion at the end of 1966 to a high of $20.9 billion by the end of 1969, and declined to $14.0 billion by the end of 1971. At the end of 1969, 46 percent of foreign commercial bank holdings of liquid dollar assets consisted of net deposits with foreign branches of U.S. banks as contrasted with 29 percent at the end of 1971. The combined liquid dollar holdings of *foreign nonbanks* rose from $11.3 billion at the end of 1966 to a high of $22.3 billion by the end of 1970, declining to $18.7 billion by the end of 1971. The variations in these holdings were due almost entirely to changes in their holdings of Eurodollars since foreign nonbank holdings of American dollar assets remained relatively stable throughout the period. Over the period 1969–71, Eurodollars represented nearly 80 percent of the total liquid dollar holdings of foreign nonbanks as contrasted with slightly over 60 percent at the end of 1966.

Over the 1966–68 period, the net Eurodollar positions of foreign nonbanks were in approximate balance. With the rapid rise in their Eurodollar deposits in 1969, their net position shifted to a credit balance of $6.3 billion by the end of that year. During 1970, foreign nonbank Eurodollar deposits leveled off, and in 1971 they declined sharply. Eurodollar loans to foreign nonbanks nearly doubled in 1970 and continued to rise during 1971. Consequently the net Eurodollar position of foreign nonbanks turned negative in 1970; the negative position increased to $10.3 billion by the end of 1971. Since the net positive position of foreign nonbanks in American liquid dollar assets remained within a range of from $3.8 to $4.8 billion over the 1966–71 period, the net combined liquid dollar position of foreign nonbanks shifted from a net positive position of $10.6 billion at the end of 1969 to a negative position of about $6.5 billion by the end of 1971.

Adequate data for determining the net Eurodollar positions of foreign commercial banks (excluding foreign branches of U.S. banks) are not available, but it seems unlikely that foreign commercial banks had a substantial *uncovered* positive dollar position in 1970 or 1971. On the other hand, foreign official institutions quite clearly had net credit positions in both Eurodollars and American dollars throughout the 1966–71 period. Although some foreign nonbanks undoubtedly incurred losses in terms of their own currencies at the time of the currency realignment in December 1971, most of them probably gained. The vast bulk of the losses from the depreciation of the dollar in 1971 were absorbed by the foreign central banks.[1]

SUBSTITUTION AMONG HOLDINGS OF DIFFERENT ASSETS

While American liquid dollar assets and Eurodollar deposits would seem to be close substitutes for one another, differentials between U.S. money market rates and Eurodollar deposit rates have at times exceeded 200 basis points. Changes in deposits of U.S. residents in the Eurodollar market are probably sensitive to changes in the U.S. money market–Eurodollar deposit rate differential, but there are simply no satisfactory data on holdings of Eurodollar deposits by U.S. residents with which to test this relationship. Our investigation of the interest sensitivity of foreign nonbanks with respect to shifts in their portfolios between interest-earning American liquid dollar assets and Eurodollar deposits failed to yield positive results. On a month-to-month basis for the period June 1968–December 1971, changes in the composition of liquid dollar

---

1. Foreign central banks provided much of the cover for the positive dollar positions of foreign commercial banks through swap operations in dollars and by supporting the forward rate on the dollar.

ssets of foreign nonbanks did not occur in accordance with changes in the spread between the Eurodollar deposit rate and the J.S. secondary market rate for CDs. In 42 observations recorded a Table 2.10, the movements were in accordance with a priori xpectations in only 10 cases. Nevertheless, better data and the nclusion of other liquid assets in the portfolios of foreign nonbanks, ogether with allowance for lags in response to changes in interest ate differentials and for changes in expectations regarding future novements in exchange rates, might well have revealed the existence f interest sensitivity. We did find indirect evidence over the period rom the end of 1968 to the end of 1970 of substitution of foreign onbank holdings of Eurodollars for their holdings of liquid American dollars. But this substitution is probably long run or structural ather than short run, that is, month-to-month or quarter-to-quarter.)ur indirect evidence of substitution is derived from the fact that uring the 1968–70 period the secular growth in foreign nonbank oldings of American liquid assets was interrupted by a decline in uch holdings. During this same period foreign nonbank holdings f Eurodollar deposits had their most rapid growth. They rose by n amount nearly twice the total volume of American liquid dollar oldings of foreign nonbanks.

Given the sharp rise after 1968 in foreign nonbank holdings of Eurodollars—all out of proportion to the historical rate of growth f their holdings of U.S. liquid dollar assets—it appears likely that nost of this increase represented a substitution of Eurodollars for ondollar currencies. There have been large differentials in yields etween Eurodollar deposits and European domestic currency deosits with the same maturity, ranging up to 300 basis points or nore on both a covered and an uncovered basis. In most cases, hese differentials have been in favor of Eurodollar deposits. While ve lack adequate data for formulating and testing a portfolio-adustment model for foreign nonbank holdings of Eurodollar deposts, our regression analysis did confirm the existence of significant elationships between changes in foreign nonbank deposits in all oreign branches of U.S. banks and in U.K. branches alone, on the one hand, and changes in differentials between covered Eurodollar leposit rates and British and Swiss deposit rates, respectively, on the other.

There is also statistical evidence that Eurodollar banks in Belgium, the Netherlands, France, and Germany over the period September 1963–June 1969 tended to be net lenders to the Eurodollar market when the interbank Eurodollar rate exceeded the domestic money market rate and to be net borrowers from the Eurodollar market when the domestic money market rate exceeded the Eurodollar rate.

EURODOLLAR MARKET AND INTERNATIONAL PAYMENTS

The operations of the Eurodollar market have a direct impact on the U.S. balance of payments only when U.S. residents are involved as net lenders or borrowers from the market. Thus during periods of large U.S. resident borrowings from the market, the U.S. *official transactions balance* tended to improve. This balance was adversely affected in periods of net U.S. resident repayments to the market. A priori analysis suggests that the growth of the Eurodollar market has had a beneficial effect on the U.S. *basic balance,* but sufficient information is not available to confirm this conclusion.

The growth of the Eurodollar market does not depend upon U.S. balance-of-payments deficits. However, to the extent that U.S. deficits increase world liquidity, foreigners have a larger volume of funds for placement in the market. Foreign official institutions have been an important source of funds for the Eurodollar market. As U.S. deficits increased the dollar reserves of foreign central banks, these institutions have been induced by the higher yields on Eurodollar deposits (and by other factors as well) to acquire Eurodollar deposits. Recently, several European surplus countries have imposed restrictions on capital imports, including the establishment of reserves on deposit liabilities to foreigners by their commercial banks and cash deposit requirements for direct borrowing by domestic nonbank corporations from the Eurocurrency markets. These actions have tended to limit the demand for Eurodollar loans in the countries employing the restrictions. The Eurodollar market and other Eurocurrency markets, however, have continued to grow partly as a consequence of the increased demand for Eurocurrency loans from the developing countries and Eastern Europe.

The International Role of the Dollar

The summary of our empirical findings provides certain insights regarding the changes in the international role of the dollar over the 1964–71 period. Foreign private holdings of American liquid dollar balances have grown only modestly when compared with the nearly twofold increase in the value of world trade and perhaps a several-fold increase in the volume of international financial transactions. Foreign nonbank holdings of American liquid dollar assets have declined since 1968. They are little higher than they were at the end of 1964; their holdings of U.S. demand deposits rose by only 13 percent between 1963 and 1971. Recorded holdings of American liquid dollar assets by foreign commercial banks (excluding foreign branches of U.S. banks) rose from $6.1 billion at the end of 1964 to $10.0 billion by the end of 1971 (Table 2.2, line 3), but of the latter amount $4.5 billion represented liabilities of U.S. agencies and branches of foreign banks to their head offices abroad, not all of which can properly be regarded as performing the functions of an international currency.

This relatively modest growth in foreign private holdings of American dollars appears to reflect two developments. First, foreigners have economized on dollar holdings which perform the functions of transactions and precautionary balances. This economy has been achieved in part through the operations of multinational banking institutions. Second, the function of the dollar as a medium of foreign private holdings of liquid interest-earning assets has in large measure been shifted to the Eurodollar market or, more broadly, to the Eurocurrency market. Nevertheless, we do not believe that these developments imply a diminution of the international currency functions of the dollar. They do imply a shift in the international financial intermediation function from U.S. financial markets to worldwide markets, a shift that occurred in part as a consequence of U.S. monetary policies, including capital export controls, and in part as a consequence of the worldwide expansion of the U.S. banking system. These two causal forces were not unrelated, of course, inasmuch as the foreign expansion of U.S.

banks was greatly stimulated by U.S. monetary policies and controls.

Our data also provide certain insights into the demand for American dollar holdings by foreign official institutions. At the end of 1964 such holdings totaled $15.8 billion, rising to $18.2 billion by the end of 1967. By the end of June 1969, however, they had fallen to $14.9 billion, mainly as a consequence of the large U.S. commercial bank borrowings from the Eurodollar market. In spite of the fact that European central bankers had been complaining about having to absorb more dollars than they wanted to hold in their reserves, in mid-1969, foreign governments complained about the drain on their dollar reserves caused by the U.S. commercial bank borrowings from the Eurodollar market,[2] and in the second half of 1969 some countries, including Germany, sold gold to the U.S. Treasury in order to rebuild their dollar reserves. This strongly suggests that, at least until the end of 1969, foreign central bank holdings of dollars were roughly in line with their voluntary demand for them. After 1969 much of the rise in foreign central bank holdings of dollars constituted an involuntary acquisition on the part of the major surplus countries of Europe and Japan.

REASONS FOR THE GROWTH OF EURODOLLARS

Although a portion of the rise in foreign private holdings of Eurodollars may be regarded as a substitute for dollar balances which might otherwise have been held in the United States, the vast bulk of the rise in these Eurodollar claims must be explained by factors relating to the growth of the Eurocurrency market itself. We may, therefore, ask whether the several-fold rise in Eurodollar deposits should be regarded as a rise in the demand for *dollar* liquidity as such or whether this rise reflected an increase in the world demand for international liquidity in general.

Over the period 1964–71, the world demand for liquid assets expanded very rapidly. In many developed countries the volume of

2. In mid-1969 when U.S. resident borrowings from the Eurodollar market substantially exceeded the U.S. dollars provided by the deficit on the U.S. basic transactions account, foreign central bankers (at the meetings of the Group of Ten) requested the United States to take action to limit U.S. bank borrowings from the market through their foreign branches. In response to this request the Federal Reserve Board established a 10 percent reserve requirement on U.S. bank borrowings through their branches beyond a specified base level.

quasi-money (time and savings accounts in commercial banks and other savings institutions) has risen proportionately more than the rise in GNP.[3] Moreover, an increasing proportion of liquid asset holdings of private nonbanking concerns and individuals appears to have taken the form of international assets. A partial indication of the rise in the volume of liquid international asset holdings is provided by the BIS data on "inside area" bank liabilities to (nonresident) nonbanking concerns and individuals. Over the period December 1966–December 1971, these liabilities rose from $4.6 billion to $12.8 billion.[4] This estimate does not include the rise of several billion dollars in foreign currency deposits held by residents of "inside area" countries with commercial banks located in their own countries. Portfolio theory offers an explanation of the expansion of demand for international assets in terms of risk diversification and the desire for higher yields on assets with equivalent maturities and risk exposure. But whatever the factors underlying the demand, the supply of international liquid assets for satisfying the preferences of portfolio holders has been provided in large measure by the development of the Eurocurrency market.

It is our view that the growth of the Eurodollar market cannot be explained in terms of either the foreign demand for dollar liquidity as such or the international liquidity requirements for an expanded international trade. Since 1964, international liquid asset holdings have increased out of all proportion to the rise in international trade. The growth of the Eurocurrency market took place in response to the increase in the world demand for diversified international liquid assets. Liquid asset holders were able to hold their assets in any one of a dozen countries with approximately the same yield for the same maturity. Thus, even though most of these liquid assets were denominated in dollars, the political risks were diversified while the exchange risks could be hedged in the forward market.

3. For example, in Germany time and savings deposits rose by nearly 190 percent over the period between the end of 1964 and the end of 1971; increasing by 130 percent in Switzerland; and by nearly fourfold in France. In none of these countries did GNP in current prices double during the period. (See Country Pages in IMF, *International Financial Statistics,* December 1971 and December 1972.)

4. BIS data include liabilities in both dollar and nondollar currencies. BIS, *Forty-Second Annual Report,* Basle, June 1972, p. 151.

THE ROLE OF THE DOLLAR IN THE EUROCURRENCY MARKET

We have noted that the Eurocurrency market has served as a very efficient international financial intermediary. First, it provides the lenders with a variety of highly liquid assets (diversified on a geographical basis) on which their covered returns are generally higher than returns available in domestic money markets, and it provides borrowers with a dependable source of loan funds with a variety of maturities and at a cost frequently lower than the cost of loans from domestic money markets. Second, the Eurocurrency market together with the Eurobond market, which have in considerable measure replaced the international intermediation functions of U.S. resident institutions, have certain advantages over U.S. resident institutions for both lenders and borrowers. Lenders can deal through their own resident banks and investment houses or through financial institutions in a number of other countries, thereby diversifying their risks. Borrowers can also deal with financial institutions in their own countries or with those in other countries. Third, both the Eurocurrency and the Eurobond markets are, to some degree at least, independent of monetary developments in the United States.[5] Nevertheless, the advantages of a worldwide system of international financial intermediation have in large measure been provided by an extension of the U.S. commercial banking and investment banking systems to the major countries of the world, working in cooperation with the financial institutions of other countries.

But why has the U.S. dollar rather than other major currencies served as the principal currency in which Eurocurrency transactions have been denominated?[6] The answer is to be found in the characteristics of the dollar itself and in the institutional framework of the Eurocurrency market. Prior to 1970 there was little expectation of

5. Assuming a continuation of reserve requirements on borrowings by U.S. banks from their foreign branches imposed by the Federal Reserve Board in June 1969 (together with subsequent Federal Reserve actions to discourage such borrowing), it appears unlikely that the heavy borrowing by U.S. banks from the Eurodollar market that took place in 1966, 1968, and 1969 will be repeated.

6. Nonbank holdings of nondollar Eurocurrency deposits were only about $0.5 billion dollars at the end of 1966, but they increased to $2.8 billion at the end of 1971 and they have continued to grow during 1972. See BIS, *Forty-Second Annual Report,* p. 151; and Morgan Guaranty Trust Company of New York, *World Financial Markets,* September 19, 1971, p. 4.

either devaluation or appreciation of the U.S. dollar vis-à-vis all other currencies taken together at least in the short run. Taking both Eurocurrency lenders and borrowers together, this made the dollar the optimal medium for international financial intermediation. No other currency had these characteristics of an international standard. In addition, the foreign branches of U.S. banks, which were largely responsible for the rapid growth of the Eurodollar market in the 1960s, sought to attract dollar deposits to provide dollars for their U.S. parents and to supply dollar loans to U.S. firms operating abroad. Multinational firms with headquarters in the United States were large depositors as well as borrowers, and they normally preferred to hold dollars. A substantial proportion of the U.S. dollars entering the Eurodollar market came from the central banks either by means of central bank deposits or through swap transactions with foreign commercial banks. The U.S. deficit on basic transactions account served to provide both a source of foreign liquidity and a source of dollars for the market during the period prior to 1970 when a large proportion of the dollars deposited with Eurodollar banks was being transferred to the United States.

Expectations of a dollar devaluation contributed to the growing U.S. deficits on official reserve transactions account during 1970 and 1971 and evidently reduced the attractiveness of Eurodollar deposits to foreign nonbank holders. Thus during 1970, foreign nonbank holdings of Eurodollar deposits rose by less than $0.5 billion (see Table 2.5) while nonbank deposits in other Eurocurrencies rose by more than $1 billion.[7] During 1971, nondollar Eurocurrency deposits of nonbanks continued to grow while Eurodollar deposits of foreign nonbanks declined.[8] This decline in foreign nonbank holdings of Eurodollar deposits was more than offset by the growth of foreign central bank deposits, of U.S. resident deposits, and of other sources of dollars available to the market (see Tables 2.3 and 2.10).

The shift in the net Eurodollar positions of foreign nonbanks

7. BIS, *Forty-Second Annual Report,* p. 151. Nonbank Eurocurrency deposits in nondollar currencies include both U.S. resident and foreign deposits, but we have no way of separating them.

8. A part of the increase in the dollar value of nondollar Eurocurrency deposits during 1971 is attributable to the depreciation of the dollar in terms of other European currencies.

from a positive balance of over $6 billion at the end of 1969 to a negative position of over $10 billion at the end of 1971 indicates that foreign nonbanks converted nearly $17 billion into foreign currencies over the period. This conversion of dollars—largely borrowed from the Eurodollar market—into foreign currencies accounted for about half of the increase in American dollar holdings of foreign official institutions over the same period. These holdings played a decisive role in precipitating the "dollar crisis" of August 1971.

Since the end of 1971, the decline in foreign nonbank deposits has been reversed. Between the end of December 1971 and the end of October 1972, foreign nonbank deposits in foreign branches of U.S. banks rose by about $1.4 billion. Deposits of foreign commercial banks in foreign branches of U.S. banks rose by about $4.7 billion over the same period, and the volume of deposits of foreign official institutions with foreign branches of U.S. banks rose also, by more than $2 billion.[9] Thus, despite the dollar crises of 1970–71, the Eurodollar market continued to expand during 1972.[10]

THE FUTURE ROLE OF THE DOLLAR

In the light of the world currency developments since August 1971 and in the context of the current negotiations on international monetary reform, two questions arise regarding the future role of the dollar. One relates to the prospective growth of foreign dollar balances in the United States. We have already witnessed retardation in the growth of foreign private holdings of American dollar balances and the partial replacement of the functions of these balances by Eurodollars. We have also seen an erosion of the international financial intermediation function of the United States. This function has in considerable measure been transferred abroad by the growth of foreign branches of U.S. banks and of U.S. investment banking houses operating in the Eurobond market. In other words, the world-banker functions of the United States have tended to become internationalized in a manner analogous to the functions

9. *Federal Reserve Bulletin,* March 1973, p. A89.
10. According to *World Financial Markets* (Morgan Guaranty Trust Company of New York, March 22, 1973, p. 4), the net size of the Eurodollar component of the Eurocurrency market grew by $15 billion during 1972.

of U.S. nonfinancial corporations. Foreign banks have also penetrated the U.S. banking system, with the result that international transactions are financed through multinational banks to an increasing degree, and international borrowing and lending take place in world markets rather than through bilateral transactions involving lenders and borrowers in different countries. It seems likely that even with a return of international confidence in the exchange stability of the dollar, international liquidity holders will continue to have a preference for Eurocurrency deposits over American liquid dollar assets and that the demand for Eurocurrency loans will continue to expand. Eurocurrency deposit rates are likely to continue to be somewhat higher than U.S. money market rates, while U.S. rates set a floor for Eurodollar deposit rates. Since U.S. capital controls played an important role in the creation of both the Eurocurrency and the Eurobond markets, it has been suggested that if these controls are phased out by the end of 1974, as Secretary Shultz proposed in his statement of February 12, 1973,[11] international borrowing will be shifted from both the Eurocurrency and Eurobond markets to the United States. However, given the institutional structure of the Eurodollar market, it appears likely that most foreign firms would find it easier and more convenient to borrow from local Eurodollar banks where they are known rather than from U.S. banks. Moreover, U.S. parent banks may prefer to make loans through their foreign branches in much the same way that they make loans through their domestic branches.

Although the efficiency of the Eurocurrency market and its popularity with both lenders and borrowers have been clearly demonstrated, we noted in Chapter 3 that national governments have been limiting their residents' use of the Eurocurrency market, and that there has been considerable discussion regarding an international agreement or concerted action by governments to constrain the operations of that market. One example of a potential general constraint on the Eurocurrency market is an agreement on the part of all governments to establish substantial reserve requirements on Eurocurrency deposits. This would increase the cost to Eurocur-

11. See "Statement on Foreign Economic Policy," Secretary of the Treasury George P. Shultz, Department of the Treasury *News Release*, February 12, 1973.

rency banks of obtaining funds from Eurocurrency deposits and would tend to reduce the interest rates the banks could pay on these deposits, or increase the rates they would have to charge on loans, or both, thereby decreasing the advantages of the Eurocurrency market as a financial intermediary. However, there are serious obstacles to the elimination or substantial curtailment of the Eurocurrency market. As an international market, the Eurocurrency market can operate anyplace in the world, beyond the jurisdiction of the major financial powers. Recently, we have witnessed the rapid expansion of Eurocurrency banking in such areas as the Bahamas, Hong Kong, Singapore, and Beirut. From all over the world, funds in large amounts can flow into the Eurocurrency market via branches of multinational banks located outside the principal financial centers, and these branches can place funds through their affiliates in almost any country. Moreover, the major developed nations are far from any agreement on the desirability of concerted action to suppress the market.

A second question is whether we should expect Eurodollar deposits to be displaced in large measure by Euro-Deutsche marks, Eurosterling, Euro-Swiss francs and other nondollar Eurocurrencies. Are the factors discussed above explaining the past dominance of the dollar in the Eurocurrency market likely to continue to prevail? The first and most important factor concerns the future role of the dollar as an international standard of value. On the one hand, this standard-of-value role has been shaken by the formal devaluation of the dollar following the Smithsonian Accord of December 1971, and by the further devaluation announced in February 1973. These events showed that the exchange value of the dollar could be changed simultaneously in relation to most of the world's leading currencies—something that many economists had been denying. Moreover, if the European Community (EC) joint float, initiated in March 1973, proves successful and is expanded to include the pound sterling, the exchange value of the dollar would fluctuate in relation to a group of the world's leading currencies. There would be two international standards of value fluctuating in relation to one another, and the consequences for the role of the dollar in the Eurocurrency market are difficult to predict. On the other hand, if the EC float does not prove successful, and if international confidence in the dollar is restored so that a further devaluation of the dollar in terms of the SDR is not anticipated, the dollar would remain as the principal international standard of value. Even in an international monetary regime in which none of the world's leading currencies maintained a formal parity (or central rate) in terms of the SDR or the dollar, the currencies of the rest of the world would be floating in relation to the dollar.[12] Under these circumstances, the dollar would remain as the major currency least likely to depreciate or appreciate simultaneously against all other currencies and, hence, would continue to be the optimal medium for international financial intermediation for both Eurocurrency lenders and borrowers.

A second factor making for the dominance of Eurodollars in the Eurocurrency market in the past has been the leading role played by U.S. banks in the Eurocurrency market in terms of both the volume of business and the provision of institutional facilities. While U.S. banks can and do trade in other Eurocurrencies, the bulk of their assets and liabilities in both the United States and abroad are in dollars, and a large part of their overseas business is with U.S.-based multinational firms and traders that deal largely in dollars. Hence, it seems likely that the most important group of banks among the "makers" of the Eurocurrency market will continue to have a strong preference for dollars. However, it has been suggested that the announced phaseout of U.S. capital controls will

12. So long as central banks intervene in the exchange market to influence the value of their currencies, a regime in which *all* major currencies are floating is virtually impossible. There will be a tendency for central banks to control the exchange value of their currency in terms of the leading international currency. Since this currency serves as the international standard of value, it cannot float in the sense that its value changes proportionately in relation to all other currencies simultaneously. This situation has been referred to as the n-1 problem. If there are n currencies, there can be only n-1 exchange rates expressed in terms of the standard of value. While the dollar is the standard, the only way that the value of the dollar can be changed by a uniform percentage in terms of all other currencies is for the value of all other currencies simultaneously to change by a like percentage in terms of the dollar. This requires an international agreement such as was negotiated in December 1971, and again in February 1973, since a uniform change in the value of all nondollar currencies in terms of the standard would almost certainly not occur as a consequence of the actions of individual countries.

mean a reduction in the operations of foreign branches of U.S. banks.[13] Nevertheless, we are inclined to believe that the penetration of the U.S. banking industry abroad, like that of U.S.-based multinational firms, is likely to remain and to grow in significance.

A third factor explaining the past position of the dollar in both the Eurocurrency and Eurobond markets is the relationship of these Euromarkets to the large U.S. financial market. To reemphasize a basic point, the Eurocurrency markets are markets for transactions *in the currencies of different countries;* Eurocurrencies are not simply units of account. The same is true of a Eurobond which calls for payment of interest and principal in dollars or in some other international currency. Eurocurrency and Eurobond operations not only involve real transfers of the currencies in which they are denominated, but large flows of funds into and out of a particular Eurocurrency market will have an impact on the domestic money market of the country whose currency is involved.

At the middle of 1972, the estimated net size of the Eurocurrency market (the BIS definition) was $85 billion, of which the Eurodollar component was $65 billion.[14] The former figure is six times the total value of Swiss franc currency and demand deposits and nearly two and one-half times the total volume of German currency and demand deposits. If the bulk of the Eurocurrency market took the form of Euro-Deutsche marks and Euro-Swiss francs—which account for over 80 percent of Eurocurrency liabilities other than Eurodollars—large movements of funds into and out of the Eurocurrency market could have very disturbing impacts on the domestic monetary systems of these countries, so much so that their monetary authorities might take action to prevent the use of their currencies for this purpose. On the other hand, large movements in and out of Eurodollars would scarcely have a noticeable impact on the U.S. money market. Moreover, the predominant position of U.S. banks in the Eurodollar market assures that a dollar "crunch" could only be of temporary duration, since the U.S. parent banks

13. See Hugh Stephenson, "Shadow Over Banks in London: American Phaseout of Curbs May Hurt," *New York Times,* February 25, 1973.

14. Morgan Guaranty Trust Company of New York, *World Financial Markets,* September 19, 1972, p. 4.

could quickly provide an ample volume of dollars.[15] The large dollar holdings of foreign official institutions also provide a potential source of support for the Eurodollar market.[16]

Implications for International Monetary Reform

At the time of writing (April 1973), negotiations are being initiated by the Committee of Twenty[17] on the reform of the international monetary system. Despite important differences among the IMF members, the addresses of the finance ministers of the major countries at the Annual Meeting of the Board of Governors of the IMF in September 1972, including that of the U.S. Secretary of the Treasury, revealed a general desire to move toward a more symmetrical role of the dollar in relation to other currencies. In his address to the Board of Governors, Secretary of the Treasury Shultz[18] outlined in broad terms a proposal for international monetary reform that would (a) reduce the role of the dollar as a reserve currency but not eliminate that role; (b) give the dollar the same technical possibilities for exchange-rate flexibility as other currencies; (c) impose on surplus countries a responsibility for balance-of-payments adjustment equivalent to that of the deficit countries; and (d) establish changes in holdings of official reserves by individual countries as the principal criterion for determining the obligations of both surplus and deficit countries to take balance-of-payments adjustment measures. A full examination of the issues raised

15. The Eurodollar market has been characterized from time to time by a temporary shortage of dollar funds which has resulted in a very sharp rise in interest rates on day-to-day funds.

16. During 1968 and 1969 when U.S. banks were borrowing heavily from the Eurodollar market, both the BIS (by drawing on its swap agreement with the United States) and European central banks intervened from time to time in the Eurodollar market to relieve the pressure on interest rates arising from heavy demands for Eurodollar funds. See, for example, Charles A. Coombs, "Treasury and Federal Reserve Foreign Exchange Operations," *Monthly Review,* Federal Reserve Bank of New York, March 1969, pp. 43–56.

17. Established at the Annual Meetings of the Board of Governors of the IMF, September 1972.

18. "Statement by George P. Shultz, Secretary of the Treasury and Governor of the Fund and Bank for the United States, at the Joint Annual Discussion," Board of Governors 1972 Annual Meetings, Washington, D.C., Press Release No. 21, September 26, 1972.

by these proposals and of alternative proposals put forward by the representatives of other IMF members at the September 1972 Annual Meeting would take us far afield. Therefore, our discussion will be limited to the implications of our analysis of foreign dollar balances for certain aspects of the international monetary reform proposals.

THE COMPOSITION OF OFFICIAL RESERVES

If the dollar is to have a reduced role in the total volume of official reserves, consideration must be given to the disposition of the large accumulated holdings of official American dollar balances, which totaled over $61 billion at the end of December 1972, and which rose to over $70 billion by mid-March 1973.[19] Foreign official holdings of American dollar balances plus official holdings of Eurodollars (the latter being estimated at $15–$20 billion at the end of 1972)[20] constituted about half of all foreign official reserves as of December 1972.[21] Our earlier analysis suggests that even with the restoration of confidence in the exchange value of the dollar, a substantial portion of the official holdings of American dollars is not likely to be shifted to foreign private holders. Increases in foreign private liquid dollar holdings are likely to take the form of Eurodollars rather than American liquid dollar assets. Some return flow of U.S. resident capital that moved into foreign currencies during 1970, 1971, and 1972 might be expected; but even if the total amount of this flow, as indicated by the errors and omissions item in the U.S. international accounts over the years 1970–72, were to return to the United States, this would take no more than $16 billion of the official dollar holdings. If we assume that this amount of short-term funds held by U.S. residents abroad returned to the United States, and if, in addition, we assume that another $10 billion were more or less permanently held by foreign official institutions as working balances, this would still leave a balance of some $35 billion in foreign official holdings of American

19. *Federal Reserve Bulletin,* March 1973, p. A78, Table 6.

20. Based on an estimate of $25–$30 billion for total Eurocurrency holdings of foreign official institutions given in *World Financial Markets,* Morgan Guaranty Trust Company of New York, March 22, 1973, p. 6.

21. *International Financial Statistics,* April 1973, p. 19.

dollars, plus whatever net accumulation takes place after December 1972.[22]

The fact that foreign central banks hold a portion of their official reserves in the form of Eurocurrency deposits also raises problems for the composition and volume of the world's official reserves. One of the purposes of the SDR facility created in 1969 (and a major objective of virtually all current proposals for international monetary reform) is the achievement of international control over the volume of official reserves. The redepositing of official reserves in Eurocurrency banks makes for instability in the volume of official reserves.[23] Such redepositing may also be employed to conceal the volume of these reserves.[24] This practice would interfere with an internationally supervised balance-of-payments adjustment mechanism based on changes in the level of each country's official reserves, such as that envisaged by the U.S. Secretary of the Treasury in the reform proposals noted above.

Regarding the Eurocurrency element in official reserve holdings, the suggestion has been made that central banks should agree not to hold Eurocurrencies. As has been mentioned earlier, in the spring of 1971, the central banks of the Group of Ten entered into an agreement not to increase their Eurocurrency holdings. While an agreement not to hold Eurocurrencies might be reached by the central banks of the major industrial countries, a large portion of the Eurocurrency holdings of foreign central banks is held by central banks in countries outside Western Europe. At the end of 1971,

22. In the first quarter of 1973, the United States had an official settlements deficit roughly estimated at $10.5 billion, seasonally adjusted. (*World Financial Markets,* Morgan Guaranty Trust Company of New York, April 24, 1973, p. 5.) A substantial portion of this deficit was probably caused by the outflow of U.S. resident capital, and most of this capital is likely to return with the advent of foreign-exchange stability.

23. As has been explained in Chapter 3, the depositing of reserve currencies in Eurocurrency banks tends to increase the total volume of the world's official reserves. Moreover, they can be quickly liquidated; and under a system of convertible currencies, the actual currencies could be presented to the monetary authorities for conversion into reserve assets such as gold or SDRs.

24. For example, at the end of October 1972, Japan reported official reserves of $17.8 billion. However, it has been disclosed that at that time, the Japanese government held $1.8 billion in deposits with foreign commercial banks, $2.5 billion in deposits with Japanese commercial banks, and $900 million in medium- and long-term foreign bonds, none of which was included in the reported holdings of official reserves. (*Wall Street Journal,* November 1, 1972, p. 6)

nearly two-thirds of the $10 billion of (identified) Eurodollar holdings of central banks were held by central banks outside the ten major industrial countries.[25] (This is probably a considerable underestimate, since, as noted in Chapter 2, there is in existence a substantial sum of unidentified Eurocurrency holdings of official institutions.) At the end of 1972, total central bank and government holdings of Eurocurrencies were estimated to be between $25 and $30 billion,[26] the bulk of which is believed to be held outside Western Europe.[27] It would be difficult, if not impossible, to convince most countries that they should hold their official reserves in SDRs or other assets yielding a low return or none at all. It is our view that Eurocurrencies are likely to become increasingly important as official reserve assets, and we see little likelihood of this trend being reversed.

DOLLAR CONVERTIBILITY AND OFFICIAL DOLLAR HOLDINGS

Most governments, including the United States, seem to envisage an eventual restoration of the convertibility of the dollar (and of other major currencies as well) into a noncurrency reserve asset, probably SDRs. Dollar convertibility will, of course, require an effective balance-of-payments adjustment mechanism plus adequate U.S. reserves to deal with temporary drains. In addition, it is generally believed that some disposition must be made of the large official holdings of American dollars as a condition for restoring and maintaining the convertibility of the dollar. There are several proposals for dealing with these official dollar holdings, none of which is entirely satisfactory to all concerned. For example, these holdings could be funded into long-term U.S. obligations with an SDR-value guarantee, or they could be exchanged at the IMF for SDRs or special IMF deposits. In either case, the tendering of the official dollar assets (or other reserve currencies) could be made compulsory or optional for the holders. However, it appears unlikely that the U.S. government would favor a compulsory arrange-

25. *IMF Annual Report 1972,* Washington, D.C., p. 30.
26. *World Financial Markets,* Morgan Guaranty Trust Company of New York, March 22, 1973, p. 6.
27. An increasing amount of the foreign-exchange earnings of the governments of the Middle East petroleum-producing countries is believed to be flowing into the Eurocurrency market.

ment in the light of its stated position that the dollar should continue to serve as an official reserve medium. Moreover, some countries might want to maintain the existing level of their reserves, while others would prefer to exchange all of, or a portion of, their reserve currency holdings for long-term bonds with an SDR-value guarantee. One proposal is that foreign central banks be given an option of exchanging their dollars for SDRs by a specified date, after which time their dollar holdings would be ineligible for such conversion. However, an arrangement of this sort is likely to create difficulties, especially if foreign countries elect to retain a large proportion of their official dollars.[28] For one thing, the availability of these official dollars for payments to the United States would mean that this country could not earn reserve assets when it had a surplus. Also, a country electing to hold its reserves in dollars might sell the dollars to buy a third currency, with the dollars flowing into other central banks. Would these dollars be convertible into SDRs, or would there be two kinds of foreign-held dollars, one convertible and the other inconvertible? A similar problem would arise if the country electing to hold dollars decided to place a portion of them in the Eurodollar market, in which case the dollars might flow into other central banks. If these dollars were convertible, the central banks acquiring them might present them for redemption to the U.S. Treasury. This could occur even when the United States was in equilibrium on basic balance account.

Even though Eurodollars are not obligations of the United States, developments in the Eurodollar market could cause a temporary drain, or threat of a drain, on U.S. reserves. A clear case would be the withdrawal by foreign central banks of a portion of the $10 to $15 billion in official Eurodollar deposits currently held and the presentation of these dollars to the U.S. Treasury for redemption in SDRs. Let us assume that a foreign central bank withdraws a billion dollars in deposits from foreign branches of U.S. banks. The immediate effect would be a transfer of American dollars to the

---

28. Many countries are likely to find American dollar assets or Eurodollars more attractive than SDRs as reserve assets. The interest yield on SDRs is likely to be much less than that on liquid dollar assets and, in addition, some countries may feel that dollars or other reserve currencies entail less risk than credits on the books of the IMF—credits which have value only so long as IMF members are willing to provide their own currencies against such credits.

foreign central bank. This action would tend to reduce Eurodollar lending by the U.S. foreign branches (or by the Eurobanking system as a whole). An equivalent amount of American dollars would, in turn, be received by the U.S. foreign branches as net Eurodollar lending declined. Some, not necessarily all, of the dollars representing net loan repayments to the U.S. branch banks might come from foreign central banks. Some of the dollar repayments might come from U.S. residents if foreign affiliates of U.S. corporations were forced to curtail their borrowing from the Eurodollar market. An increase in Eurodollar deposit rates might attract additional deposits from other central banks, unless there was an expectation of a depreciation in the value of the dollar. The rise in interest rates would also attract funds from the United States. Much the same chain of events would occur if the foreign central banks withdrew Eurodollar deposits from foreign commercial banks instead of U.S. foreign branches. The foreign commercial banks would either draw down their balances held in the United States or, perhaps more likely, would reduce their Eurodollar deposits with foreign branches of U.S. banks. In either case, there would be a transfer of American dollars to foreign central banks and, thus, an addition to official holdings eligible for conversion.

Developments along the foregoing lines could be set off if expectations of some future U.S. devaluation were to trigger a massive reduction of Eurodollar deposits, both private and official. The immediate effect would be a flow of American dollars to foreign central banks. Again, as Eurodollar lending was curtailed, there would be a reflux of dollars to the foreign branches of U.S. banks, part of which would come from the foreign central banks as foreign borrowers converted their domestic currencies into dollars to make repayments. During 1971, when there was a substantial decline in foreign nonbank Eurodollar deposits, it was offset in part by a rise in foreign official deposits, in part by dollars acquired by foreign commercial banks through swap arrangements with central banks, and in part by a flow of U.S. dollars to the market. The first two sources might not be readily available if foreign central banks had the option of acquiring other reserve assets from the United States. The third source would simply add to the flow of dollars to foreign central banks.

It should be emphasized that these are short-term drains that would be reversed as long as there was no long-run shift in the total foreign demand for American liquid dollar assets, or a substantial outflow of U.S. private funds to the Eurodollar market. Basically, what would be involved is a short-term capital flow from the U.S. parent banks to deal with a liquidity crisis in the Eurodollar market. Nevertheless, in view of the huge volume of Eurodollar claims, such flows could be quite sizable. The possible magnitude of these flows is difficult to estimate, but they could amount to many billions of dollars, quite apart from short-term capital exports by U.S. non-banking residents to the Eurodollar market.

Clearly, contingencies like those discussed in the preceding paragraphs would need to be taken into account in determining the level of U.S. reserves compatible with the convertibility of the dollar into SDRs, and in defining the extent of such convertibility with respect to accumulated balances. In the latter regard, a possible compromise solution would be to set some minimum percentage, adjusted to the circumstances of each country, for the amount of its official dollar claims (whether held directly or as Eurodollars) to be converted into long-term obligations.

THE ROLE OF THE DOLLAR AS AN INTERVENTION CURRENCY

Another key issue in any international monetary reform program concerns the role of the dollar as an intervention currency. Throughout the postwar period, the dollar has had no rival for the intervention function. Following the Smithsonian Accord of December 18, 1971, most major industrial countries adopted "central rates" expressed in terms of the U.S. dollar, and agreed to maintain their exchange rates within a margin of 2¼ percent above and below the central rate for their currency registered with the IMF, giving a total spread of 4½ percent in relation to the dollar. Other countries, including France and the United Kingdom, maintained their existing par values with the Fund, while others adopted new par values. This system of par values, or central rates, was weakened somewhat by the British decision to float the pound sterling in June 1972, and the system collapsed in February and March of 1973. The second devaluation of the dollar, by approximately

10 percent, was announced on February 19, 1973, together with the floating of the Japanese yen. Intense speculation against the dollar and certain other currencies led to the initiation in March 1973 of a joint float among the currencies of the following EC countries: Germany, France, the Netherlands, Belgium, Luxembourg, and Denmark; they were later joined in the float by the non-EC countries of Sweden, Norway, and Finland. The currencies of the other EC countries—the United Kingdom, Ireland, and Italy—floated independently of the EC float but were expected to become stabilized in relation to the other EC currencies at some time in the future. The joint float involves the maintenance of exchange rates among the EC currencies within a maximum margin of 2¼ percent, but there is no fixed trading relationship with respect to the dollar. The EC currencies in the joint float are maintained within the 2¼ percent band in relation to each other by means of intervention conducted in those currencies, while intervention in dollars apparently takes place unilaterally, with the central bank of each country influencing the relationship of its currency, and hence that of the entire group, to the dollar.

Should the EC joint float, enlarged eventually by the pound sterling and the Italian lira and perhaps other currencies, prove successful and enduring, another currency—possibly the Deutsche mark—will share the intervention role with the dollar. In time, there may be developed a common EC monetary unit which is convertible at a fixed rate into each EC currency, and which will be traded on the international currency markets.

At the present time, both international financial instability and the serious negotiations being conducted by the Committee of Twenty for the reconstitution of the international monetary system make it difficult, and perhaps foolhardy, to predict the future of the dollar as an intervention currency and international standard of value. Nevertheless, we believe it would be unwise to write off these functions of the dollar for the future, at least until another currency emerges with the prerequisites for performing them. These prerequisites explain the dominant international role of the dollar over the past several decades, namely, a large stake in world trade, investment, and banking, and a massive volume of international obligations denominated in the national currency. Perhaps only a common EC monetary unit such as that described above could meet these qualifications. The fact that a currency is strong does not necessarily make it a good candidate for the standard of value and intervention role. In fact, a very strong currency is quite likely to appreciate vis-à-vis all other currencies and, hence, would be unacceptable to international debtors.[29] Conversely, a very weak currency becomes unacceptable to international creditors.

It is quite possible that the partial floating rate system in effect since February 1973 might continue with the dollar serving as the principal intervention currency. Under this system, major foreign countries are intervening in the foreign-exchange market to control the dollar value of their currencies but at the same time are avoiding any substantial additional accumulation of American dollars. This may well prove to be the most feasible means of achieving and maintaining a pattern of exchange rates consistent with general balance-of-payments equilibrium. However, the fact that virtually all IMF members favor a return to some form of par value system in which currency parities are expressed in terms of the SDR[30] suggests that in time the world may abandon the present floating rate system and adopt a par value system based on the SDR. But the SDR in its present form cannot be privately held or traded in the exchange markets against national currencies, nor does it seem likely that such an SDR will be devised. Consequently, the SDR cannot serve as an intervention currency. Moreover, a par value system based on the SDR cannot be established until an intervention currency exists which is stable in terms of the SDR and convertible into it. The fact that the dollar is formally defined in terms of the SDR but is not convertible into it means that the conditions for an enduring par value system are not met.

29. The European Payments Union was established in 1950 precisely because the dollar was too strong to serve as the intervention currency for freely convertible European currencies. De facto convertibility among the European currencies was achieved via the Clearing Union.

30. Nearly all of the statements made by the Governors of the IMF at the Board of Governors 1972 Annual Meeting favored a par value system based on the SDR. Only France among the major industrial countries favored defining parities in terms of gold and the establishment of an international gold standard system.

The above analysis suggests that if the dollar is to perform twin roles of intervention currency and de facto standard of value in a future par value system, there must be a restoration of confidence in the stability of the dollar in terms of the de jure standard, the SDR, and this would seem to require convertibility of the dollar into SDRs. Furthermore, the U.S. objective of providing full symmetry between the dollar and other major currencies so that the dollar has the same degree of exchange flexibility as other currencies, even to the extent of enabling the dollar temporarily to float freely on the exchange markets, would appear to be incompatible with the intervention role.[31] While under a new international monetary system there may occur changes in the value of the dollar in terms of the SDR, such changes must necessarily be infrequent as compared with parity changes of other currencies. Frequent changes in the value of the dollar in terms of the SDR might well lead to chaotic conditions in the exchange market, since a large number of countries would tend to follow the dollar while others would maintain their currency parities in terms of SDRs, thereby changing the entire pattern of cross rates. If and when the EC countries are able to evolve a common monetary unit, and this unit gains prominence both as a transactions medium and an intervention currency among a large number of countries, the principal focus in a decision to change the value of the dollar in terms of SDRs would, in effect, be a change in the relationship between the dollar and the EC monetary unit, with most of the currencies of the rest of the world lining up with one or the other of the two de facto standards.

The problems sketched above have important implications for exchange-rate adjustments, which are expected to play a far more important role in balance-of-payments adjustments in the future than in the past. While we do not regard these problems as insoluble, outlining possible solutions, such as a system of multiple currency intervention or a system under which the IMF would regulate

31. For example, if the United States wanted the dollar to float freely temporarily, it could do so only if all the major industrial countries would agree not to employ the dollar as an intervention currency, i.e., not to buy or sell dollars against their own currencies in the exchange market. Exchange rates between nondollar currencies would need to be maintained within the prescribed limits above and below parity by the use of a nondollar intervention currency.

currency parities in order to maintain effective exchange rates,[32] is beyond the terms of reference of this study.

32. The effective exchange rate is the weighted composite value of all other currencies in terms of a country's currency. Practically, it has significance only in terms of a percentage change in the effective rate from one period to another. For an excellent discussion of the proposal for achieving exchange-rate adjustments by means of altering effective rates—thereby solving the problem of changing the value of the currency that serves as the intervention currency—see Henry C. Wallich, *The Monetary Crisis of 1971—The Lessons to Be Learned* (Per Jacobsson Foundation Lecture, September 14, 1972), Washington, D.C.: International Monetary Fund, 1972.

Glossary

The following glossary of terms is confined to certain terms that have a special meaning as employed in this volume. It does not cover terms with which students of international finance are generally familiar.

Foreign dollar balances include all liquid claims of foreigners on U.S. residents, plus liquid dollar claims held by nonresidents of the United States on other nonresidents.

Foreign liquid claims on U.S. residents include all claims designated as liquid by the Department of Commerce in the U.S. balance-of-payments tables published in the *Survey of Current Business,* plus a few additional claims set forth in Appendix A of Chapter 2, above.

Foreign liquid dollar claims on nonresidents of the United States are confined to Eurodollars.

American dollars are liquid claims on U.S. residents.

Eurodollars are dollar-denominated time deposits in commercial banks outside the United States, including foreign branches of U.S. banks.

U.S. liquid claims on foreigners include all such claims designated as liquid by the U.S. Department of Commerce in the balance-of-payments tables published in the *Survey of Current Business,* plus certain additional claims set forth in Appendix A of Chapter 2 of this volume.

Foreign commercial banks are commercial banks located outside the United States, excluding foreign branches of U.S. banks.

Foreign nonbanks are private individuals and concerns, other than commercial banks, not regarded as residents of the United States for balance-of-payments purposes.

Foreign official institutions are central banks and governmental authorities and their agents located outside the United States.

Multinational banks are U.S. and foreign commercial banks having branches and agencies in countries other than that in which their head office is located.

Eurocurrencies are time deposits in banks outside the United States, including foreign branches of U.S. banks, denominated in a currency other than that of the country in which the bank is located.

"Inside area" consists of those countries for which data on commercial banking operations are reported regularly to the BIS. The countries

114

are Belgium, France, Germany, Italy, the Netherlands, Switzerland, Sweden, and the United Kingdom.

"Outside area" consists of the rest of the world not included in the eight countries listed above.

"Inside area" banks are commercial banks in "inside area" countries, including foreign branches of U.S. banks plus the BIS (the assets of which are included in the data recorded for the Swiss commercial banks).

"Inside area" nonbanks are private individuals and concerns other than commercial banks, located in the countries that comprise the "inside area."

"Inside area" nonresident nonbanks are private individuals and concerns, other than commercial banks, located in "inside area" countries but not in the country in which the "inside area" bank with which they have a deposit or from which they have received a loan is located.

Eurobanks are all commercial banks, wherever located, that receive deposits in currencies of countries other than that in which the bank is located.

Inter-Eurobank dollar market is the market for dollars among Eurobanks in the worldwide redepositing system.

"Bank" sources are sources of dollar funds to "inside area" banks arising from "inside area" central banks and the BIS and from dollars purchased by "inside area" banks.

Appendix Tables

APPENDIX TABLE 1

Foreign Liquid Claims on the United States, 1950–57
(millions of dollars; end of period)

Date	Short-Term Liabilities Reported by Banks to:			Long-Term Government Securities[a]	Total
	Official	Commercial	Private		
1950	3,620	2,100	1,390	1,280	8,390
1951	3,550	2,600	1,510	610	8,270
1952	4,650	2,630	1,680	900	9,860
1953	5,670	2,570	1,780	810	10,830
1954	6,770	2,570	1,810	750	11,900
1955	6,950	2,980	1,790	1,310	13,030
1956	8,040	3,410	2,030	1,100	14,580
1957	7,920	3,470	2,250	1,220	14,860

SOURCE: IMF, *International Financial Statistics, Supplement to 1964–1965 Issues*, p. 250.
a. Includes holdings of international organizations.

116

APPENDIX TABLE 2

U.S. Liabilities to Foreign Central Banks and Governments, 1957–72[a]
(millions of dollars; end of period)

End of Period	Demand Deposits	Short-Term Time Deposits	Short-Term U.S. Govt. Obligations	Acceptances and Other Short-Term	Marketable U.S. Treasury Bonds and Notes	Long-Term Bank Deposits	Non-marketable Treasury Obligations	Total
1957	—3,059—		4,246	612	775			8,692
1958	—3,511—		4,392	762	610			9,275
1959	—2,834—		5,738	582	966			10,120
1960	—3,037—		6,193	982	876			11,088
1961	—3,389—		6,458	1,092	890			11,830
1962	—3,234—		7,807	922	751		200	12,914
1963	1,402	2,451	7,578	1,036	1,183	9	766	14,425
1964	1,591	2,816	7,554	1,259	1,125	158	1,283	15,786
1965	1,535	2,862	7,186	1,483	1,105	120	1,535	15,826
1966	1,679	2,668	6,832	1,360	860	913	584	14,896
1967	2,054	2,458	8,137	1,378	908	1,807	1,452	18,194
1968	2,149	1,899	5,949	1,321	462	2,341	3,219	17,340
1969	1,930	2,942	3,844	2,361	346	1,505	3,070	15,998
1970	1,652	2,554	13,367	1,760	295	695	3,452	23,775
1971	1,327	2,039	32,311	3,341	1,955	144	9,534	50,651
1972p	1,583	2,858	31,448	4,052	5,501	94	15,747	61,284

SOURCES: *Federal Reserve Bulletin*, various issues; U.S. Department of the Treasury.
p = preliminary
a. Includes Bank for International Settlements and European Fund. Short-term liabilities to the BIS and European Fund reported by U.S. banks constitute the vast bulk of the short-term liabilities to "Other Western Europe" reported regularly in the *Federal Reserve Bulletin* and in the *Treasury Bulletin*. These amounts are as follows (millions of dollars; end of period):

1961—325	1965—369	1969—1,553
1962—351	1966—234	1970— 594
1963—465	1967—706	1971—1,391
1964—358	1968—357	1972—1,483

118 *Appendix Tables*

APPENDIX TABLE 3

Liquid Claims of Foreign Commercial Banks on U.S. Residents, 1957–72

(millions of dollars; end of period)

End of Period	Demand Deposits	Short-Term Time Deposits	Long-Term Time Deposits	Short-Term U.S. Govt. Obligations	Acceptances[a] and Other Short-Term[b]	Roosa Bonds	Total	U.S. Bank Liabilities to Foreign Branches[c]
1957	—2,737—			158	577		3,472	n.a.
1958	—3,017—			131	372		3,520	n.a.
1959	—3,369—			492	817		4,678	n.a.
1960	—4,210—			76	532		4,818	n.a.
1961	—4,917—			43	524		5,484	n.a.
1962	—4,678—			83	585		5,346	n.a.
1963	4,102	838		68	808		5,816	793 (1-1-64)
1964	5,027	967		142	1,167		7,303	1,183
1965	4,941	900		113	1,465		7,419	1,345
1966	6,636	1,243	25	137	1,921		9,962	4,036
1967	7,763	1,142	15	129	2,051		11,100	4,241
1968	10,374	1,273	8	30	2,794	125	14,604	5,598
1969	16,756	1,999	55	20	4,870	135	23,835	12,561
1970	12,385	1,354	166	14	3,417	135	17,471	6,218
1971[d]	3,400	320	257	8	7,223	135	11,343	1,276
1972p	4,673	546	259	5	9,630	135	15,248	1,454

SOURCES: *Federal Reserve Bulletin*, *Survey of Current Business* and U.S. Department of the Treasury.

p = preliminary

a. Includes negotiable certificates of deposit.

b. Includes items payable in foreign currencies. Includes portions of liabilities to non-banks payable in foreign currencies.

c. Included in total but distribution among categories of liquid claims cannot be identified.

d. There was an important change in coverage of the official statistics in 1971: (a) liabilities of U.S. banks to their foreign branches and those liabilities of U.S. agencies and branches of foreign banks to their head offices and foreign branches, which were previously reported as deposits, are included in "Other Short-Term Liabilities"; (b) certain accounts previously classified as "Official Institutions" are included in "Banks"; and (c) a number of reporting banks are included in the series for the first time.

APPENDIX TABLE 4

Liquid Claims of Foreign Nonbanking Concerns and Individuals on U.S. Residents, 1957–72
(millions of dollars; end of period)

End of Period	Demand Deposits	Short-Term Time Deposits	Long-Term Time Deposits	Short-Term U.S. Govt. Obligations	Acceptances and Other Short-Term[a]	Marketable U.S. Treasury Bonds and Notes[b]	Total
1957	—1,766—		n.a.	278	209	445	2,698
1958	—1,951—		n.a.	306	174	375	2,806
1959	—1,833—		n.a.	295	270	541	2,939
1960	—1,849—		n.a.	148	233	550	2,780
1961	—1,978—		n.a.	149	230	516	2,873
1962	—2,096—		n.a.	116	352	448	3,012
1963	1,493	966	19[b]	119	469	341	3,407
1964	1,531	1,271	46[b]	72	503	376	3,799
1965	1,574	1,594	79[b]	87	332	472	4,138
1966	1,513	1,819	50	83	329	528	4,323
1967	1,693	2,054	40	81	297	558	4,723
1968	1,797	2,199	40	86	362	465	4,989
1969	1,711	1,935	40	107	312	525	4,630
1970	1,688	1,895	53	131	325	565	4,655
1971	1,660	1,666	56	96	271	447	4,196
1972p	1,954	2,025	87	65	481	425	5,037

SOURCES: *Federal Reserve Bulletin* and U.S. Department of the Treasury.

p = preliminary

a. Claims payable in foreign currencies included in foreign commercial bank claims—Appendix Table 3.

b. Includes some holdings of foreign commercial banks.

120 *Appendix Tables*

APPENDIX TABLE 5

U.S. Liquid Claims on Foreign Official Institutions, 1957–72
(millions of dollars; end of period)

Period	Short-Term Loans Reported by U.S. Banks	U.S. Official Currency Holdings	Total
1957	242	n.a.	242
1958	401	n.a.	401
1959	351	n.a.	351
1960	290	n.a.	290
1961	329	116	445
1962	359	99	458
1963	186	212	398
1964	221	432	653
1965	271	781	1,052
1966	256	1,321	1,577
1967	306	2,345	2,651
1968	247	3,528	3,775
1969	262	2,781	3,043
1970	119	629	748
1971	224	276	500
1972p	166	241	407

SOURCES: *Federal Reserve Bulletin*, various issues; *Treasury Bulletin*, various issues.
p = preliminary

APPENDIX TABLE 6

U.S. Liquid Claims on Foreign Commercial Banks, 1957–72
(millions of dollars; end of period)

| Period | Reported by U.S. Banks | | | | Reported by U.S. Nonbank Concerns | | Total |
	Short-Term Loans	Acceptances Made for Account of Foreigners	Deposits in Foreign Currencies	Other	Deposits in Dollars	Deposits in Foreign Currencies	
1957	386	a	132	699[b]	n.a.	n.a.	1,217[c]
1958	440	a	181	656[b]	n.a.	n.a.	1,277[c]
1959	498	a	203	582[b]	n.a.	n.a.	1,283[c]
1960	524	a	242	1,233[b]	n.a.	n.a.	1,999[c]
1961	709	a	386	1,874[b]	n.a.	n.a.	2,969[c]
1962	953	a	371	1,967[b]	640	—	3,931
1963	955	2,214	432	384	640	—	4,625
1964	1,403	2,621	336	803	1,060	—	6,223
1965	1,567	2,508	329	492	700	—	5,596
1966	1,739	2,540	241	464	757	109	5,850
1967	1,616	3,013	287	467	852	128	6,363
1968	1,697	2,854	336	509	1,219	272	6,887
1969	1,943	3,202	352	670	1,062	183	7,412
1970	1,720	3,985	352	766	697	173	7,693
1971	2,080	4,254	548	1,686	1,075	234	9,877
1972p	2,976	3,215	441	2,478	n.a.	n.a.	n.a.

SOURCES: *Federal Reserve Bulletin*, various issues; *Treasury Bulletin*, various issues.
p = preliminary
a. Included in "other."
b. Includes acceptances made for account of foreigners.
c. Claims on foreign commercial banks by U.S. nonbank concerns were not reported, and hence are not included in the totals.

APPENDIX TABLE 7

U.S. Liquid Claims on Foreign Nonbanking Concerns, 1957–72
(millions of dollars; end of period)

Period	Reported by U.S. Banks[a]	Reported by U.S. Nonbanking Concerns[a]	Total
1957	15[b]	n.a.	n.a.
1958	16[b]	n.a.	n.a.
1959	15[b]	n.a.	n.a.
1960	238[b]	n.a.	n.a.
1961	200[b]	n.a.	n.a.
1962	186[b]	n.a.	n.a.
1963	157	n.a.	n.a.
1964	187	379	566
1965	68	145	213
1966	70	107	177
1967	70	182	252
1968	40	147	187
1969	89	247	336
1970	92	271	363
1971	173	195	368
1972p	223	n.a.	n.a.

SOURCES: *Federal Reserve Bulletin*, various issues; *Treasury Bulletin*, various issues.
p = preliminary
a. Includes foreign government obligations and commercial and financial paper, all payable in foreign currencies.
b. Includes "other short-term claims" payable in foreign currencies.

Index

U.S. liquid claims on foreigners (Cont.)
and net U.S. liquidity position, 17–20, 87

Voluntary Foreign Credit Restriction Program (VFCR)
effects on U.S. basic balance, 82

and growth of Eurodollar market, 82–83

Wallich, Henry C., 113
Willms, Manfred, 84
Woolley, Herbert B., x